AGRICULTURAL VALUE CHAIN

AGRICULTURAL VALUE CHAIN

Henry Mayo
Kevin Longbow

Kruger Brentt
Publishers

2025

Kruger Brentt Publishers UK. LTD.
Company Number 9728962

Regd. Office: 68 St Margarets Road, Edgware, Middlesex HA8 9UU

ISBN: 978-1-78715-279-3

For information on all our publications visit our website at http://krugerbrentt.com/

PREFACE

Nestled within the intricate tapestry of global food systems, the agricultural value chain stands as a linchpin, intricately weaving together the efforts of farmers, producers, distributors, retailers, and consumers to ensure the flow of food from field to table. This interconnected web of activities encompasses not only the physical processes of cultivation, harvesting, processing, and distribution but also the myriad economic, social, and environmental factors that shape the production and consumption of food.

The significance of the agricultural value chain extends far beyond mere sustenance; it underpins economic development, social cohesion, and environmental stewardship on a global scale. As the world's population continues to grow, urbanize, and diversify, the pressures on agricultural value chains intensify, driving the need for innovation, efficiency, and sustainability throughout the system.

At the heart of this scholarly exploration lies a deep-seated curiosity about the inner workings of agricultural value chains and the myriad forces that govern their operation. From the rural landscapes where crops are sown and livestock are raised to the bustling urban markets where food is bought and sold, the agricultural value chain encompasses a vast and diverse array of actors, activities, and relationships.

Through this interdisciplinary inquiry, we seek to illuminate the complexities inherent in the agricultural value chain, examining not only its structural components and functional dynamics but also its broader socio-economic implications. By drawing upon insights from fields as diverse as economics, sociology, agronomy, and supply chain management, we aim to unravel the intricate web of interactions that shape the performance and sustainability of agricultural value chains around the world.

In embarking on this scholarly odyssey, we are guided by a shared commitment to advancing knowledge, fostering dialogue, and catalyzing positive change within the realm of agricultural value chains. By shining a light on the challenges and opportunities facing this critical component of our global food system, we hope to inspire informed action and innovative solutions that promote food security, economic prosperity, and environmental resilience for generations to come.

Henry Mayo
Kevin Longbow

CONTENTS

01 UTILIZING THE AGRICULTURE VALUE CHAIN TO ENHANCE INCOME DISTRIBUTION: A CASE STUDY OF INDONESIA

1. INTRODUCTION: AN OVERVIEW

Globalization provides emerging countries with opportunity to grow their businesses in domestic and international markets. Agriculture is a crucial tool for sustainable development and alleviating poverty [1]. Hence, the producer must vigorously compete with other producers and establish a competitive advantage by enhancing the product's worth. To accomplish these objectives, it is essential to regulate and manage the activities across the supply chain of the agricultural commodity, starting from the plantation farm to the store. The agriculture value chain oversees the movement of products and information throughout the supply chain, capturing the increased value at each stage. It provides a chance to lower the expenses and uncertainties throughout the supply chain. Moreover, the perishability of agricultural commodities requires specific material handling and treatment to maintain material quality. Various agricultural commodities likely require distinct material handling and processing methods. In recent times, food safety and food security have become worldwide concerns due to increased awareness and concern for health. These conditions incentivize manufacturers to enhance their awareness and ensure the quality of the product for the customer. Each stakeholder should manage operations throughout the agricultural product supply chain to reduce costs and risks while enhancing customer-perceived value.

Indonesia is a growing country with significant potential in agricultural goods. Indonesia's geographical circumstances result in abundant natural resources, particularly in agriculture products such as horticulture, fish, and livestock. Indonesia's incapacity to manage resources is a significant limitation. Vegetables are one of the valuable horticultural commodities in Indonesia. According to Ref. [2], the vegetable production in Indonesia reached 11,005,954 tonnes/year for the last 5 years. The national vegetables production also increases every year, but this increase of production should be followed by good quality of vegetables. The Indonesian agro-industry must maximize the utilization of natural resources by enhancing value throughout the supply chain. An analysis of the current state of various agricultural products' value chains can help the author pinpoint hurdles and hazards, enabling them to take suitable actions to address these challenges in the industry. Efficiently managing and monitoring supply chain activity will optimize the value chain and reduce logistical costs. Insufficient infrastructure and poor coordination in the supply chain result in rising costs and ineffective material and information transfer. Collaboration among supply chain players, strategic approaches, and government backing are essential to enhance the implementation of the agriculture value chain in Indonesia.

This article provides an overview of the agriculture value chain in Indonesia for several commodities including seafood, aquaculture, and horticulture. Case studies on supply chain risk management (SCRM) and analysis of logistics activities are used to assess and enhance the agricultural potential in Indonesia. SCRM assists in pinpointing potential risks in each level and offering suitable mitigation strategies to minimize the impact of each risk. Additionally, doing a study of logistical activity within the supply chain using activity-based pricing is necessary to assess and enhance the agriculture value chain. The studies utilized convenience sampling as the sampling method and conducted data collection through in-depth interviews with each respondent along the supply chain. The sample size in each case study varies.

2. THEORETICAL FRAMEWORK

Ref. [3] explains that value chain analysis (VCA) aims to understand how a business generates value for consumers by evaluating the impact of various activities within the organization on the overall value. A value chain involves interconnected activities that provide value for the customer. Productivity, efficiency, and the complexity of agricultural value chains are crucial factors that support commercial agriculture and agribusiness [4].

Key obstacles for producers in underdeveloped countries include the absence of a conducive environment providing institutional and physical assistance, resource availability, and efficient coordination in value chains. Small-scale producers are disadvantaged due to limited capital, reliance on traditional processes, dependence on family labor, and lack of communication with international market participants. [5, 6, 7].

Two complimentary methods can be used to promote agro-enterprise growth for competitiveness and involvement. To enhance the investment environment and attract private investors, especially small and medium-sized enterprises (SMEs). Surveys conducted in Indonesia, Nicaragua, Sri Lanka, and Tanzania show that the scarcity of rural financing, infrastructure, business, and public services is a significant constraint. The alternative strategy focuses on addressing obstacles in the growth of small and medium-sized agricultural businesses, specifically within value chains. [4].

According to Ref. [8], factor conditions relate to the nation's endowment with resources such as physical, human, knowledge, technology and infrastructure. These factors enable or constrain value chain upgrading. Moreover, Ref. [9] revealed that three key elements for a balanced analysis of value chain are network structure, horizontal and (vertical) market channel relationships, value added and governance. Moreover, value chain actors may be motivated to improve their position in the chain by changing their production of value added, their relationships (governance) with other actors in the value chain and by choosing different market channels for their products.

SCRM is an attempt to control the risks in the supply chain of a commodity which provides recommendations for the stakeholders to optimize the supply chain activity. Each tier along the supply chain may handle different kinds of risks in which the activities to prevent the risks in each tier might be different as well [10]. The practice of SCRM consists of identifying the risk from the risk owner, analysing the probability and severity of the risk, monitoring the risk and continuously evaluating the risk.

In most value chains, each activity has a distinct cost structure determined by different cost drivers. Analysing cost requires disaggregating the value chain to identify the relative importance of each activity with respect to total cost, the cost drivers for each activity, how cost in one activity influence the others and which activities should be undertaken or outsourced [11].

Several main activities related to logistics are the design of supply chain, procurement or purchasing, transportation, receiving, warehousing, material handling, distribution, return, replacement, disposal of waste and communications. The most important thing in an activity of logistics is how to make the whole stakeholder in the supply chain working together to obtain an efficient material flow [12].

Calculation of logistics costs in every activity can be used to determine which tier has a dominant proportion of cost against activity [13, 14, 15]. The logistics activity along the corn supply chain is divided into six activities including procurement, material handling, maintenance, transportation and communication. The logistics cost analysis shows the proportion of logistics cost and its component cost, the most influence activity on their respective logistics activities and what activities can be controlled at every tier [16, 17].

From Ref. [18], one underlying aspect to appreciate in terms of customer demand or usage requirements is the relationship between the customer order-to-fulfilment lead-time (CLT) and the sum of the supplier order-to-fulfilment lead-time (SLT), the firm's cycle time (CT) and the delivery-to-customer lead-time (DTC) as shown in Figure 1. CLT denotes the amount of time a customer is willing to wait, once an order has been placed, to be satisfied by the firm; SLT denotes the amount of time the firm is willing to wait for its own wishes to be met by its suppliers in producing what the customer wants once the customer order is received; CT denotes the amount of time it takes the firm to manufacture and process a customer order and finally, DTC denotes the amount of time it takes the firm to deliver a completed customer order to the customer, so that CLT = SLT + CT + DTC.

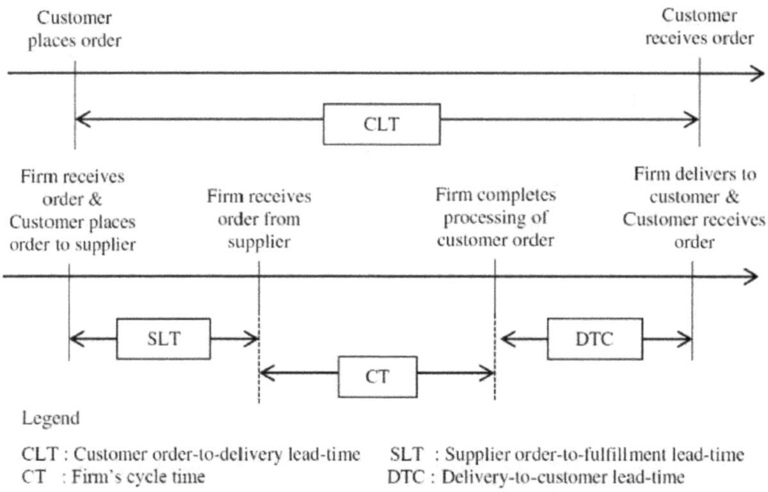

Figure 1. Lead time approach for inventory classification.

3. INDONESIAN AGRICULTURAL VALUE CHAIN

Indonesia's agriculture industry holds significant potential for economic growth due to the country's rich biodiversity, which serves as both a strength and a problem in global competition. Indonesia's strong consumptive culture should be shifted towards more productive industries to enhance the country's competitiveness globally. Japan, China, and Thailand have been concentrating on agricultural growth to improve their competitive edge. One issue in Indonesian agriculture is the failure to enhance the value of products before selling them. Indonesia sometimes exports raw commodities at a low cost. Upon reaching the target country, the raw material is enhanced through processing to create a completed product, which is subsequently imported to Indonesia at an increased cost. It is definitely a disadvantage for Indonesia. On a national level, primary suppliers such as farmers and fishermen provide the least value to the commodity. It leads to a reduction in their profits and places them in a position where they have no bargaining power and must accept market prices. The absence of oversight and support from the local government impacts the distribution of income within a supply chain.

Agro-industry in Indonesia is experiencing tremendous growth across small, medium, and large scales. Agro-industry in Indonesia plays a crucial role in enhancing the value of agricultural products by engaging in food processing. The notion of the agriculture value chain can be utilized on a small scale for various commodities like horticulture, rice, and fisheries, in addition to large-scale production. Generating income in agriculture entails not only marketing the produce but also engaging in activities such as tourism, sales, processing, seedling, education, and more. Increasing the value of the commodity might also raise the price of the product. Hence, the stakeholders throughout the value chain might generate more profits.

4. ORIGINAL CASE STUDY

Several case studies in Indonesia focus on applying value chain analysis to various agricultural commodities inside the agriculture value chain practice.

4.1. Freshly Caught Sea Fish Case

Indonesia is a growing country with significant potential in agricultural goods. Indonesia's geographical circumstances result in abundant natural resources, particularly in agriculture, including seafood. Indonesia produced 6,105,225 tonnes of fish, including 5,707,012 tonnes of fresh marine fish and 398,213 tonnes of aquaculture fish. Approximately 1,081,717 tons of fish caught in Indonesia were sourced from the coastal region of Java Island [19]. Catchment fish as perishable products requires special material handling during distribution and storage activity. Therefore, each stakeholder along the supply chain should control the quality of product through the temperature control. Based on Figure 2, the production of catchment fish significantly increases each year. In 2009, the production of catchment fish reached 4812 thousand tonnes, then it increased to 5039 thousand tonnes in 2010 and 5346 thousand tonnes in 2011. The increase of production did not significantly occurred in 2012 which only reached 1.6%, while the increase in 2013 reached 5%.

The growing population of fish caught in a specific area suggests the future promise of this resource. Currently, the rise in manufacturing quantity is not being accompanied by adequate material handling procedures, potentially resulting in a decline in product quality. Quality degradation could affect the product's selling price and result in financial losses. Cold supply chain management is essential to limit quality deterioration throughout distribution by keeping the temperature of fresh fish below 0°C to reduce microbial development.

Furthermore, there are six main tiers in the supply chain of catchment fish including fishermen, ship owner, fish auction facility (FAF), trader, fish processing units (FPUs) and end customer. Data from 64 respondents consisting of 22 fishermen, 22 ship owners, 8 FAFs, 9 traders and 11 fish processing units (FPUs) were collected from the coastal area of Java Island. Fishermen in this study are the fishermen who do not have their own ships so they rent a ship on vessel providers (ship owner) and make payment based on the share-fishing. Based on this condition, fishermen will be the stakeholder who gets the greatest loss if the selling price is cut down due to the quality deterioration. Furthermore, Ref. [20] revealed that fishermen have low bargaining power in term of price, so the government provides fish auction facility (FAF) as an infrastructure to enhance the productivity and improve the price of catchment fish in the fishermen stage. FAF assists fishermen by providing services related to fish landing, auctioning, processing, and marketing through ship owners and traders. Both the ship owner and merchant are required to pay punishment, with the specific sum varying for each Free Alongside Ship (FAF) agreement. It constitutes around 5% of the whole catchment fish productivity.

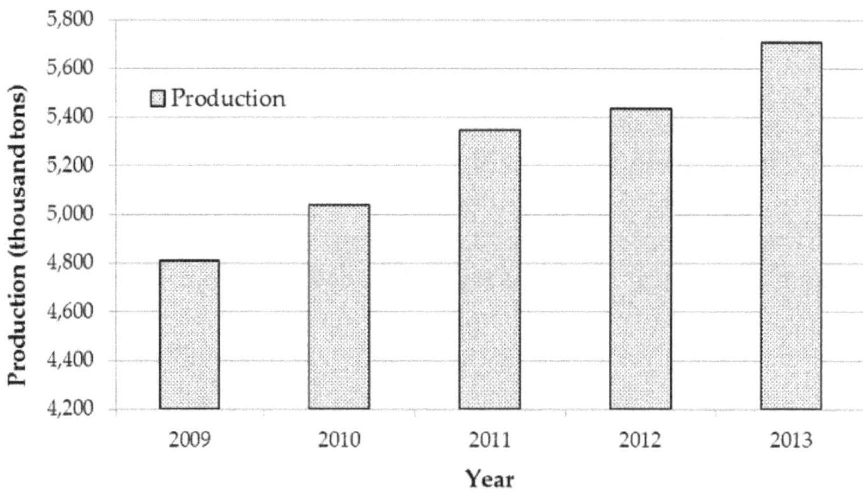

Figure 2. The production of fresh catchment sea-fish in Indonesia [19].

Enhance the value added in each tier by optimizing every activity in the supply chain. The present catchment fish supply chain is not operating optimally due to insufficient infrastructure and inefficient practices, as revealed in a thorough interview. To enhance the value added in the supply chain, it is essential to identify the most useful activity so that improvements can be made to gain a competitive edge.

4.1.1. Logistics Cost of Fresh Catchment Sea-Fish

In the practice of cold supply chain, every logistics activity in each tier needs particular cost to encourage the appropriate temperature for controlling the quality of product. However, in some cases, the large cost required becomes the constraint in the cold chain practice. The total cost along the cold supply chain in a whole cannot be separated by any logistics activity at each tier. In the integrated supply chain system, changing in costs of particular logistics activity in one tier may affect others and the whole supply chain. The analysis of the logistics cost along the supply chain of catchment fish thus should be taken to identify the proportion of logistics costs, determine the cost components of the most influential and valuable activity and define which activities can be controlled to minimize the overall logistics costs. In addition, the logistics cost analysis can be used as one of the considerations in determining the selling price and profit margin.

Moreover, fish that had completed the auction process was then placed in a container, such as Styrofoam or cold box, and ice cube was added to keep the product at the low temperature. Transportation used for distribution is differentiated based on the distribution distance where truck, container or car is used for long distance trip and motor cycle is used for short distance trip. In the peak season, the catchment fish which are not distributed yet will be stored in a particular period by trader. Wholesalers will keep them in the cold storage, while the small traders will keep in the cold box which is filled with ice. According to Ref. [21], temperature of cold storage which is around from −18 to −30°C may freeze the fish product by turning almost all the water content in the product into ice because the temperature is under the freezing point of water. Therefore, the use of cold storage tends to be more optimal to control the quality of fish.

The logistics activity in this case study encompasses procurement, material handling, inventory, transportation, maintenance and information. Calculation of logistics cost is only carried out on the *Euthynnusaffinis* as the most dominant type of catchment fish in this study. In addition, Figure 3 shows the proportion of logistics cost for non-freezer (using the ship with the cold storage and longer fishing trip) and freezer chains (using ice cube for storage during the trip and shorter fishing trip). Based on Figure 2, transportation cost accounts for the largest portion of the logistics cost with the slight difference between freezer (43.57%) and non-freezer chains (42.91%). It is because the freezer chain requires more fuel for longer trip and cold storage. For both chains, transportation cost of trader is dominant because traders deliver large quantity of product with limited capacity of vehicle which leads to the high intensity of delivery. On the other hand, procurement accounts for the second largest portion and it is only owned by the fishermen. Procurement cost includes needs for fishing and labour cost which is entirely covered by fishermen. The material handling, maintenance, information and inventory costs have little portion in the logistics cost in which total of those four accounts for less than 25%.

Material handling costs are mostly controlled by ship owners and traders. These costs consist of post-fishing expenses, inspection fees, tool depreciation, and losses incurred during handling. The post-fishing cost includes reprisal costs incurred during the auction process in the FAF, with ship owners and traders being responsible participants. Fishermen mostly contribute to the maintenance cost of tools and vehicles, whereas traders and ship owners mainly incur information costs such as communication bills. Inventory costs are minimal because to the fast-moving nature of catchment fish, making inventory management a less significant effort compared to other logistical operations.

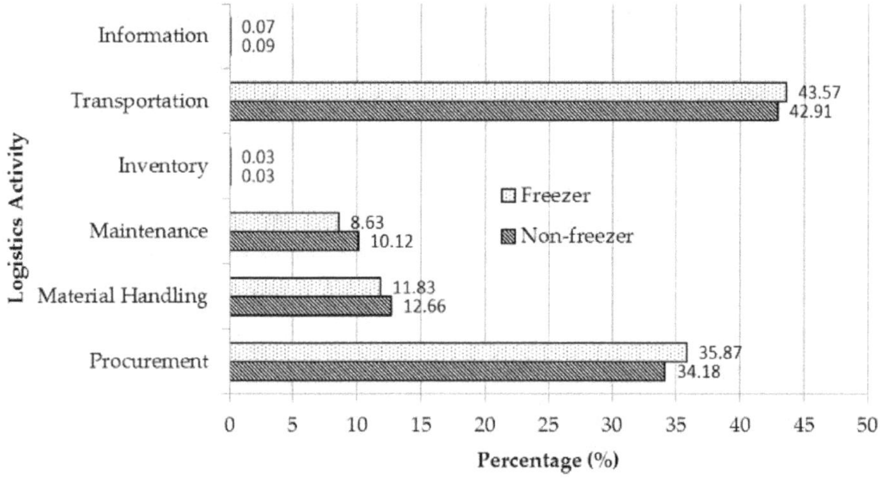

Figure 3. The proportion of logistics cost for each logistics activity.

The selling price of catchment fish in the market fluctuates significantly, ranging from Rp 9000 to 14,000/kg in the fishermen stage and Rp 13,000 to 23,000/kg in the trader stage. Based on the current market price, the market price of catchment fish processed by trader is sold at average price Rp 17,125/kg, while catchment fish in the fishermen stage is sold at a price Rp 11,375/kg in average as shown in Figure 4. The large gap occurs in the ship owner stage because they get the fish from the share-fishing with the higher portion and spend less operational cost. It is very difficult for the fishermen to get profitable margin since they are responsible for all the operational cost during fishing and share the fishing result with the ship owner. The fishermen play a role as a price taker because they have low bargaining power, while the ship owners have higher bargaining power in the auction process. With the material handling and information cost ranging from Rp 664 to 882/kg, the ship owners then sell the fish at a ranging price from Rp 11,125 to 12,625/kg. Total margins in the whole value chain are Rp 5750 (freezer) and 6625/kg (non-freezer) with the highest portion in the ship owner stage. Therefore, in terms of making profit, the highest profit margin ratio is gained by the ship owner among the value chain players. The high profit margin shows the good achievement for each stakeholder, but when looking at the whole value chain, unbalance profit margin leads to inefficiency value chain.

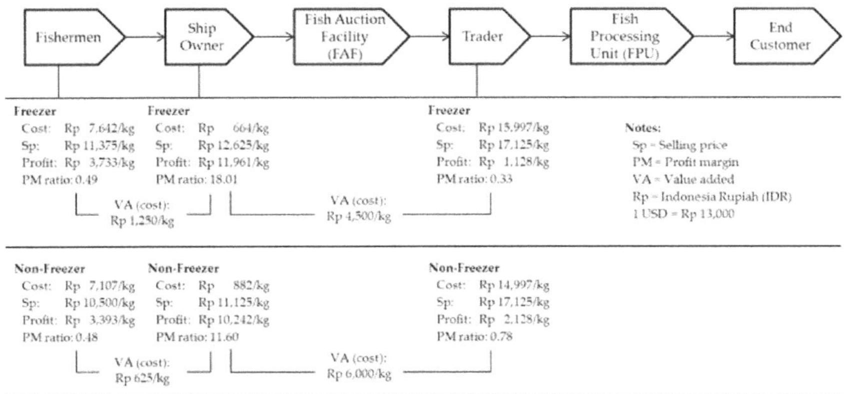

Figure 4. Value chain of fresh-sea fish.

4.1.2. Formulating Supply Chain Strategies

Improving the value chain efficiency of the fish catchment can be achieved by optimizing the profit margin ratio at each level. One option is to transition from a share-fishing system to a lease or rental system where rental costs are determined by the duration of the rental and the amount of fish caught by fishermen. Furthermore, the growing demand for catchment fish as a functional product presents both an opportunity and a challenge for stakeholders throughout the value chain. Nevertheless, the ability to provide appropriate quantity and good quality of products with profitable margins for every stakeholder is challenging. According to Ref. [22], the implementation of efficient strategy is the appropriate strategy for functional product where it focuses on the fulfilment of customer demand at the lowest prices by reducing the total cost of the activity with the largest cost. This case study focuses on reducing costs at the fishermen and trader phases due to the ship owners' decreased expense. Fishermen can lower their procurement costs by utilizing an alternative equipment known as 'rumpon' to decrease diesel fuel use and by participating in an association to collectively address procurement expenses. The government should offer sustainable assistance and training to fishermen, closely monitor selling prices to ensure all stakeholders earn profitable margins, and invest in infrastructure like affordable cold storage to support the cold chain system. Cold storage improves the quality of the product and allows fishermen to sell it at a greater price. On the other hand, trader should consider the Full Truck Load to minimize the transportation cost per unit.

4.2. Case of Fresh Vegetables

Vegetables are a significant horticultural commodity in Indonesia. National vegetable production increases annually and should be accompanied by high-quality veggies. Vegetables require specialized management throughout the supply chain due to their perishable nature. The research was carried out in the vegetable center located in the highland areas of the Magelang region, Central Java, and the Sleman region, Yogyakarta. Figure 5 displays the three primary stages involved in the distribution of fresh vegetables.

Due to their perishable nature and fast-moving classification, vegetables do not need to be stored for extended periods, resulting in lower inventory costs across the supply chain. This research revealed that inventory cost of fresh vegetables accounted for 1.22% of total logistics cost which is equivalent to Rp 28.58/kg. Inventory cost comprises of holding cost, labour cost and losses during inventory. In addition, Figure 5 shows the increasing portion of inventory cost along the supply chain. Farmer accounts for only 2.14% of the whole inventory cost, while trader has the largest portion of inventory cost in the whole supply chain. Trader incurs higher holding costs, including electricity and space leasing, compared to collector and farmer, due to managing a bigger volume of items. The storage duration lengthens as the supply chain progresses from the collector and trader obtaining fresh veggies from several suppliers to distributing them to customers.

In order to enhance this value chain, both trader and collector should control their inventory activity by strengthening the integration and communication in terms of quantity demand among the stakeholder. Therefore, the inventory can be efficiently managed.

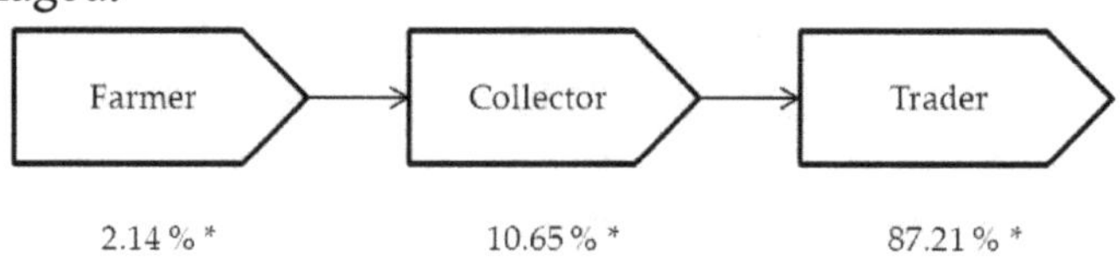

*) Percentage from total inventory cost in the whole supply chain

Figure 5. The supply chain of fresh vegetables and inventory cost percentage in each tier.

In other cases, the study of lead time in the inventory management of fresh vegetables was conducted in the Yogyakarta region. The data were collected based on purposive sampling method, followed by in-depth interviews and group discussions to suppliers at each tier. The results from Ref. [15] show that several vegetable varieties have positive lead time (CLT > SLT + CT + DTC) as characterized by predictable demand and relatively the consumer preferences change rapidly as shown in Table 1. This type of supply chain is reliable and predictable in terms of delivery quantities. In this case, supplier has a strong bargaining power, because they have many suppliers (more than 50 farmers). Approximately 80% of the varieties are non-unique plant such as cabbage, squash, leeks, pack coy, chicory, tomatoes and small cherry tomatoes.

Meanwhile, 20% of the rest (other farmers) grows unique varieties, namely broccoli, lettuce, green lettuce, beetroot, red spinach and kaylan. These vegetables have negative lead time (CLT < SLT + CT + DTC) as indicated by demand level that is difficult to be predicted but relatively occupy stable consumer preferences. The lead time in this supply chain tends unreliable due to the fluctuation in the quantity of delivery. The bargaining power of suppliers is weak as it is only supplied by a limited number of farmers.

Based on this fact, the most proper inventory management for non-unique vegetables is inventory postponement, while the most proper inventory management for non-unique vegetables is inventory speculation.

4.3. Aquaculture Business Case

During year 2010–2014, aquaculture production showed a positive trends with an increasing reach 23.74%/year on average The positive performance for the value of aquaculture production also increases in the same period with an average increase of 16.12%/year. Meanwhile, demand from Java Island will continue to grow because per capita consumption of fish in Java Island is still below per capita consumption outside Java Island. This case study was conducted in the Sleman region and the Klaten region, while Tilapia (*Oreochromis niloticus*) and Catfish (*Siluriformes*) are in the focus of this case study.

The main stakeholders in supply chain of aquaculture consist of six tiers including fish larvae producer, spreader, enlarger, wholesaler (collectors), retailer and consumer. In a supply chain, the amount of value added provided by each tier could be different because of their different respective functions in the activities of the supply chain running. In the analysis of logistics cost, Ref. [22] revealed that procurement activity is accounted for the highest portion in the total logistics cost of Tilapia and Catfish in aquaculture business. Procurement activity as the initial activity in this aquaculture business is very important because it will determine the success of the rest of the activities in this business. Good initial activity leads to the good result and yield. In addition, the second highest cost is material handling.

Margin value is obtained from the difference between selling price and purchasing price. Figure 6 shows that the highest margin occurs in the spreader stage, while the lowest one is in the enlarger stage. Spreader gives value added by producing Tilapia with the size equal to 50–60 pcs/kg. Different margins between the Sleman region and the Klaten region are influenced by differences in the production yield of Tilapia fish in each area which may affect the selling price of the fish. Moreover, distribution margin for Catfish between the Sleman and Klaten regions is not significantly different (Figure 7) due to the similarity of commodity price in both areas. Similar with Tilapia fish, the highest margin occurs in the spreader stage. On the other hand, the lowest margin in the value chain of Catfish is in the collector stage.

Table 1. Management inventory at supplier level.

	Vegetables	Lead-time (hours)				Supply chain characteristics	Inventory management
		CLT	CT	SLT	DLT		
Non-unique vegetables	Cabbage	16	4	3	6	Predicted demand, fluctuative consumer's demand, reliable supply chain, predicted delivered products' quantity, strong bargaining power, many suppliers	Speculation
	Squash	48	4	6	6		
	Scallions	12	2	2	6		
	Pakcoy	12	2	2	4		
	Chicory	12	4	4	4		
	Tomato	24	4	6	6		
	Small tomato	24	4	6	6		
Unique vegetables	Broccoli	12	6	6	6	Unpredicted demand, stable consumers' demand, unreliable supply chain, unpredicted delivered products' quantity, weak bargaining power, few suppliers	Postponement
	Green lettuce	8	4	4	4		
	Lettuce	8	4	4	4		
	Beetroot	12	8	6	6		
	Red spinach	8	4	4	4		
	Kailan	8	4	4	4		

Notes: CLT: customer order-to-delivery lead-time; SLT: supplier order-to-fulfilment lead-time; CT: firm's cycle time; DLT: delivery-to-customer lead-time.

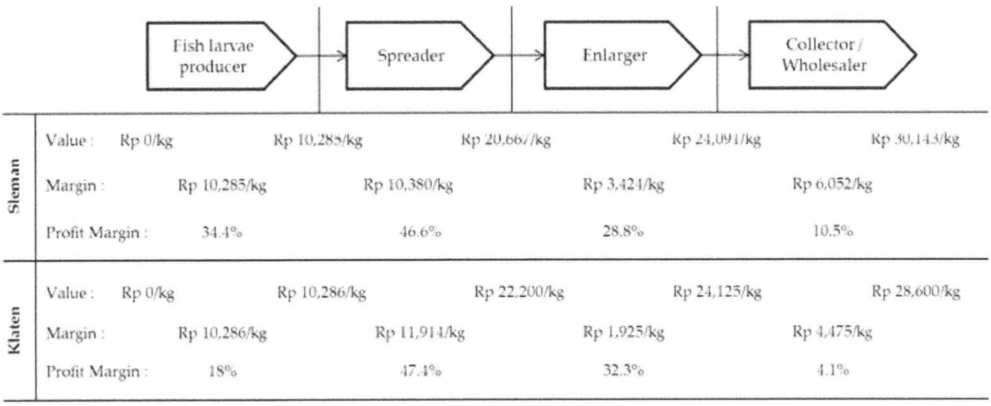

Figure 6. The value chain of Tilapia fish in the Sleman and Klaten regions.

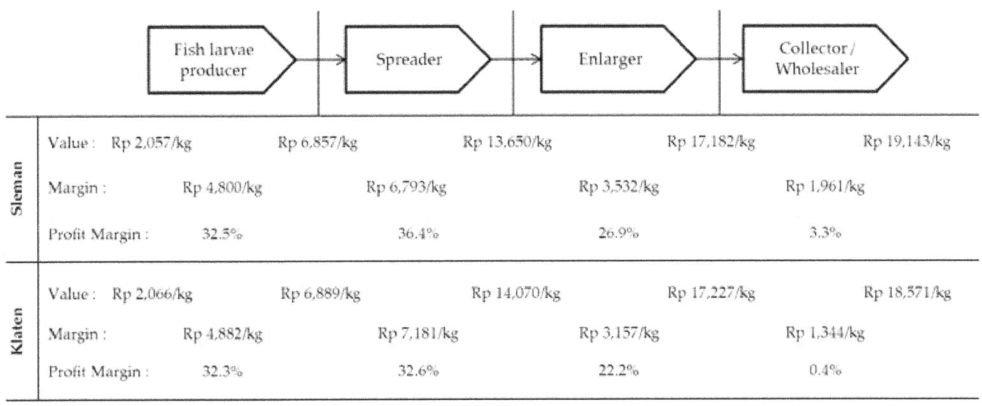

Figure 7. The value chain of Catfish in Sleman and Klaten region.

From the calculation of profit margin, different profit margins earned by each tier are occurred as shown in Figures 6 and 7. For aquaculture of Tilapia and Catfish, spreader earns for the highest profit margin among all tiers in both the Sleman and Klaten regions, while wholesaler or collector earns for the smallest. The profit margin of Tilapia fish larvae producers in the Klaten region is lower than that of producers in the Sleman region. The water temperature in the Klaten region is lower than in the Sleman region, yet the spawning process requires warm water with a temperature of roughly 25°C. This leads to increased resource utilization in the Klaten region, resulting in greater costs to achieve the same yield quantity. (see Figure 6).

The rising aquaculture production, driven by increasing demand, presents a significant opportunity for rapid growth in the aquaculture industry. To improve its profitability and sustainability, stakeholders' input should be considered. The fish larvae producer should enhance the quality of the spawning process and promote the use of cost-effective feed alternatives of equal quality, as suggested by the value chain analysis. As the intermediary, the spreader should improve the capacity to select high-quality fish seeds and choose a supplier capable of delivering superior fish seeds. Conversely, vertically integrating with the fish larvae producer could result in increased access to raw materials. The enlarger should create a cost-effective alternative feed to supplement the main feed and implement a push-based technique. To obtain a reduced price or discount, it is advisable for them to purchase in bulk from bigger businesses or suppliers. Wholesalers should implement a pull-based strategy, engage many suppliers to manage inventory, and increase sales volume. [23].

5. CONCISE SUMMARY

In Indonesia, there are many long supply chains where the role of intermediaries becomes quite dominant. In many cases, the added value e added of the product. This has led to a considerable gap between the selling prices received by the initial suppliers (farmer or fishermen) and purchasing price paid by the consumer. In addition, the integration between stakeholders has not been applied so that the profit is not optimal as a whole. Currently, creating added value can be achieved by understanding the regional resource, connecting them and disseminate the result.

In several aforementioned case studies, stakeholders who gain the highest profit are the intermediary parties with relatively low operational costs or logistics costs. Low bargaining power of initial supplier (farmer or fishermen) is still a fundamental obstacle why the profitgiven to the product in each tier of supply chain is relatively minimal, so that the increase in prices is not followed by the valu earned is sometimes not worth the effort that has been nurtured. In addition, the additional charges given to the initial supplier are sometimes too big, causing the smaller profit earned. Inequalities in profit among stakeholders show the inefficiency of supply chain. Furthermore, value chain analysis can be performed to determine which activities classified into core and support activities. Analysis using the cost parameters enables the determination of level income distribution in a commodity. The analysis can be used to formulate an improvement strategy. In this case, the government also needs to play an active role in controlling and mentoring. Therefore, the distribution of profit and the income gap among the stakeholders in a particular commodity is not too significant.

6. AT A GLANCE

Customized material handling and processing methods are necessary for individual agricultural goods.Indonesia prioritizes enhancing the productivity, competitiveness, and efficiency of the agriculture value chain to achieve competitiveness. This chapter provides an overview of the agriculture value chain in Indonesia and includes case studies on supply chain risk management and logistics cost. The author may then suggest solutions to enhance the efficiency of the agricultural value chain. Each agricultural commodity likely possesses distinct tiers, supply chain risks, challenges, and activities, resulting in varying proportions of logistics and distribution costs within each tier. The results indicated variations in approach, either speculation or deferral, for inventory management to enhance the value of horticulture within the supply chain. Similar discrepancies were observed in aquaculture research. Furthermore, the value chain analysis aids in recognizing the value generated by each party involved. In the catchment fish value chain, ship owners have a significant role in revenue distribution, whereas in aquaculture, spreaders achieve the biggest profit margin. The trader contributes the most value during shipping but receives the least profit. When analyzing the value chain of fresh vegetables, traders account for the largest part of the total inventory cost.

REFERENCES

1. World Bank. World Development Report 2008: Agriculture for Development [Internet]. 2008. Available from: http://documents.worldbank.org/curated/en/587251468175472382/World-development-report-2008-agriculture-for-development [Accessed: 17-February-2017]
2. Directorate General of Horticulture. The Production of Vegetablesin Indonesia [Internet]. 2016 [Updated: 2016-08-16]. Available from: http://www.pertanian.go.id/Data5tahun/pdf-HORTI2016/2-Produksi%20Nasional%20Sayuran.pdf [Accessed: 28 January 2017]
3. Pearce JA, Robinson RB. Strategic Management: Formulation, Implementation, and Control. 11th ed. Boston: McGraw-Hill Higher Education; 2009. p. 928. DOI: 0071263756, 9780071263757
4. Asian Development Bank. Evaluation Knowledge Study: Support for Agricultural Value Chain Development Report [Internet]. 2012. Available from: https://www.adb.org/sites/default/files/evaluation-document/35898/files/eks-agriculturalvaluechain.pdf [Accessed: 28 January 2017]
5. De Janvry A, Sadoulet E. Achieving success in rural development: Toward implementation of an integral approach. Agricultural Economics. 2005;32(s1):75-89. DOI: 10.1111/j.0169-5150.2004.00015.x
6. Daviron B, Gibbon P. Global commodity Chains and African export agriculture. Journal of Agrarian Change. 2002;2(2):137-161. DOI: 10.1111/1471-0366.00028
7. Reardon T, Timmer P, Berdegue J. The rapid rise of suoermarkets in developing countries: Induced organizational, institutional and technological change. Agrifood System. 2004;1(2):168-183

8. Porter ME. The Competitive Advantage of Nations. New York: Free Press; 1990. p. 896. DOI: 0684841479

9. Trienekens JH. Agricultural value Chains in developing countries a framework for analysis. International Food and Agribusiness Management Review. 2011;14(2):51-82

10. Guritno AD. Supply chain risk management: An approach to reduce the agricultural product's logistics costs. In: International Conference on Agro-industry (ICoA) 2015; 7-9November 2015; Japan. UEA: KnowledgeE; 2015. pp. 6-11. DOI: 10.18502/kls.v3i3.397

11. Grant RM. Contemporary Strategy Analysis. 7th ed. United Kingdom: John Wiley & Sons Ltd; 2010. p. 516. DOI: 978-0-470-95377-8

12. Waters CDJ. Supply Chain Risk Management: Vulnerability and Resilience in Logistics. United Kingdom: Kogan Page Publisher; 2007. p. 256. DOI: 0749448547, 9780749448547

13. Lailossa G. W. The new paradigm of cold chain management systems and it's logistics on Tuna fishery sector in Indonesia. AACL Bioflux. 2015;8(3):381-389. DOI: 10.1108/09600030310503334

14. Pishvaee MS, Basiri H, Sajadieh MS. National logistics cost. In: Farahani RZ, Asgari N, Davarzani H, editors. Supply Chain and Logistics in National, International and Governmental Environment. Berlin: Physica-Verlag HD; 2009. p. 4. DOI: 10.1007/ 978-3-7908-2156-7

15. Guritno AD, Fujianti R, Kusumasari D. Assessment of the supply chain factors and classification of inventory management in supplier's level of fresh vegetables. Agriculture and Agricultural Science Procedia. 2015;3:51-55

16. Gumus T, Guneri F. Multi-echelon inventory management in supply chain with uncertain demand and lead time: Literature review from an operational research perspective. Proceedings of the Institution of Mechanical Engineers, Part B: Journal of Engineering Manufacture. 2007;221:1553-1570

17. Lailossa GW. The new paradigm of cold chain management systems and it's logistics on tuna fishery sector in Indonesia. AACL Bioflux. 2015;8:381-389

18. Wallin C, Rungtusanatham MJ, Rabinovich E. What is the "Right" inventory management approach for a purchased item? International Journal of Operations and Production Management. 2006;26(1):50-68. DOI: 10.1108/01443570610637012

19. Statistics Indonesia. Production of Fresh Sea-Fish According to Province and Sub-sector (ton) on 2000 – 2013 [Internet]. Available from: www.bps.go.id [Accessed: 19-April-2016]

20. Guritno AD, Suwondo E. An application of data envelopment analysis to determine the efficiency level of the fish auction facility in coastal area, java Island. In: Proceeding of 1st International Conference on Tropical Agriculture (ICTA 2016); 25-26 October 2016; Yogyakarta. New York: Springer International Publishing; 2017

21. FAO. Freezing and Refrigerated Storage in Fisheries [Internet]. Available from: http://www.fao.org/3/contents/19deb470-de18-5d73-92ce-a2caefdac8ea/v3630e08.htm [Accessed: 20 February 2017]

22. Fisher ML. What is the right supply chain for your products? Harvard Business Review. 1997;75(2):105-116

23. Guritno AD. Logistics cost structure analysis for the development of supply chain strategies on aquaculture business. In: Proceeding of The Asian Business and Management Conference (ABMC); 13-15 October 2016; Japan. Japan: The International Academic Forum (IAFOR); 2016. p. 29-34

02 SAMENMARKT® PROPOSES TO RESTORE TRUST IN THE HORTICULTURAL FRESH FOOD MARKET THROUGH THE UTILIZATION OF MULTI-AGENT SYSTEM TECHNOLOGY

1. INTRODUCTION: AN OVERVIEW

In recent decades, Dutch greenhouse vegetable supply chains have been consistently changing. Until 1996, producers owned their local auction cooperatives cooperatively. The auction clock set the daily price for produce, which was agreed upon collectively and all transactions were processed through it. These legal cooperative enterprises joined into larger regional collectives to enhance efficiency, market position, and lower expenses. As the market evolved, the demand for decentralized auctions also shifted. In 1996, all collectives except Veiling ZON joined into a single cooperative organization called VerenigdeTuinbouwveiling Nederland (VTN), to supply produce weekly as desired by retail. The growers were represented by VTN's commercial affiliate, The Greenery BV. The Greenery, holding almost 90% of the Dutch supply market, aimed to establish itself as the primary supply chain partner for Dutch vegetables by directly selling to retailers without involving external trade parties. This goal was pursued by acquiring ownership of three major wholesale companies in 1998. The market share of The Greenery has decreased from 90% to around 20% for several product groupings throughout the years.

In the transition to The Greenery, the choice was made to change the market mechanisms significantly. The known and trusted auction mechanism, the clock, was replaced by a much less transparent mechanism, namely negotiation of deals directly with retailers, negotiated by traders employed by The Greenery. Growers used to the auction clock as a fair and just pricing mechanism did not trust the prices set by The Greenery's traders. Within the first few years, growers started to leave The Greenery. They set up new cooperative organizations to organize the sales of their produce. Today instead of the two organizations that were active in

1996, Veiling ZON and The Greenery, there are now six cooperative organizations that represent the growers.

During that time, a few growers' companies have merged and grown. In 1996, the largest grower had approximately 10 hectares of glasshouse, but today, a few producers have more than 50 hectares. Scaling up necessitates substantial financial inputs, typically obtained through loans from local banks. Increased volume leads to higher profits. The growth was facilitated by the promising markets in the emerging Central European countries, but the credit crises in 2008 drastically altered the market dynamics. As prices dropped, more growers had severe financial difficulties with less assistance from banks. Traders caused extreme market volatility through speculating, manipulating pricing of virtual goods, giving the illusion of excess supply and driving prices down. In 2015, about 65% of the growers were facing difficulties in surviving due to this condition [1, 2].

Since 2008, several initiatives have been implemented to convert the supply chain into a more logical market-focused system. Financial constraints have significantly impacted the cultivators' ability to be flexible, creative, and entrepreneurial. They mostly concentrate on price. This also applies to other parties in the supply chain, such as cooperatives and traders. There is significant distrust between these parties, hindering their ability to develop a plan for change for their future. To thrive in the horticulture fresh food market, stakeholders in the supply chain must rebuild confidence. To rebuild trust, all stakeholders must be involved in the change process to create sustainable survival and well-being for each of them.

This chapter introduces the concept of SamenMarkt®, a distributed market concept designed to re-establish trust between stakeholders in the food supply network using a multi-agent system-based Internet technology to coordinate demand and supply, supporting multiple distributed market mechanisms to support the production of horticultural fresh food in relation to price negotiation, provisioning and distribution of the produced products.

Section 2 utilizes the YUTPA framework to analyze the current supply chain dynamics and determine the necessary features for the SamenMarkt® platform. The YUTPA framework is utilized to pinpoint trust elements concerning survival and well-being. Section 3 outlines the advantages of using a multi-agent system strategy for designing systems in scattered markets within a supply chain. Section 4 elaborates on the notion of SamenMarkt®. Section 5 offers a preliminary overview of the research design and plan for implementing changes. Section 6 examines the advantages and disadvantages of the SamenMarkt® proposition.

2. EXAMINING THE PRESENT SITUATION USING THE YUTPA FRAMEWORK TO FIND TRUST-RELATED ELEMENTS IMPACTING SURVIVAL AND WELL-BEING.

Partners in the supply chain are unable to collaborate due to a lack of confidence, hindering the development of a shared plan for transforming the horticulture fresh food market into a more sustainable organization. This chapter provides a YUTPA analysis, which is described as 'being with You in Unity of Time, Place, and Action,' on how trust is built in relation to well-being and survival for each stakeholder in the market. [3]. The YUTPA framework distinguishes four dimensions of place, time, relation and action, each with their own different factors of significance. In a specific situation, a factor has specific values. The values are determined in discussions with stakeholders by posing questions on the importance and value of each factor in specific situations [4]. Trade-offs for building trust are converted into design specifications for SamenMarkt®. [5].

YUTPA assessments are conducted for the primary stakeholders in the horticultural fresh food market, including growers, cooperatives, traders, retailers, and consumers[6]. Below, the discussion covers the compromises related to trust across four dimensions – place, time, action, and relation – for various stakeholders (grower, trader, retailer, and consumer) in the horticulture fresh food supply chain network. The prerequisites for SamenMarkt® are displayed in italics.

2.1. Place

- **Body sense:** From consumers' perspective, the quality of food affects the experience of taste, the amount of household waste [7], as well as the contribution of food to their health [8]. From growers' perspective, there is a direct relation between the quality of the food and their personal effort. Many growers work long days and are unsure of the quality of their produce till the very last moment. In addition, they do not know the price for which they will be able to sell their produce, creating uncertainty and stress. Traders and retailers do not have a direct physical relation to the horticultural fresh food produce. They strive for the best quality at the lowest possible price.

SamenMarkt® needs to represent the physicality of the horticultural fresh food market and to address the sense of quality of the products.

- **Emotional space:** Growers are proud of their companies. Their physical location demands continual attention with many factors of uncertainty to the value of their work (weather, quality, prices and timing). The emotional space is defined by uncertainty on the one side and pride and autonomy on the other side. For traders, the emotional space is defined in a business environment in which all strive to win and get better quality for less money. It is a 'game-environment' where results of activities are only measured in financial gain. Some retailers, like the growers, are proud of their business. However, the vast majority of retail consists of supermarkets in which marketing and management all strive for efficient and successful business. The main driver for retailers is the necessity to have all products with the proper specifications available every day of the year, no matter how high the demand is, at the lowest price or at most the same price the competitor is paying. For consumers, the emotional space is defined by the availability of products, the price, the appearance, the experience of shopping, the conformity with their life style and the experience of taste.

SamenMarkt®must contribute to a more balanced and relaxed climate in the supply chain network by taking away uncertainties and strengthening ties within the supply chain network. Strengthening ties create understanding for other stakeholders' positions and increases awareness of interdependencies in the supply chain network. Individual acceleration needs to be tempered by the context of the whole network.

- **Environmental impact**: For growers, a sustainable environmental impact is very important. It defines the value and quality of their products. The weather, the land, fertility and possible pollution are all factors of importance to growers. Traders, retailers and consumers are rarely confronted with environmental impact. They expect the grower to deal with these aspects.

SamenMarkt®should offer insight and respect for the growers' practice in dealing with environmental impact.

- **Situated agency**: In the current horticultural fresh food market in the Netherlands, it seems that only the traders and the retailers have the sole power to influence the supply chain network [9]. They rule negotiation processes and are free, for example, to create virtual produce to lower prices. Consumers can only buy or not buy, growers produce as much as possible. The communication between consumers and producers is blocked and by 2015 completely dependent on the price negotiation between traders and retailers who do not offer transparency, not even to the growers.

SamenMarkt® should provide greater autonomy to growers inside the supply chain network. Growers themselves must be equipped and supported for this task. Retail firms closely analyze consumer engagement to predict future behavior based on past purchasing patterns. Consumer agency, encompassing awareness and responsibility, can be improved. Price is determined by traders and retailers. They should enhance transparency regarding their actions, pricing, reasoning, and quality of service.

2.2. Time

- **Duration of engagement**: For retailers and traders, the duration of engagement is very different to the duration of engagement of the growers and to the duration of the engagement of the consumers. Currently, as the produce travels through the supply chain network, engagement stops when it enters a next step in the supply chain network, i.e. as the result of change of ownership. Growers have no insight into what happens after a trader has bought the goods. Retailers are unaware of what growers do; consumers are unaware of how quality relates to price. Duration of engagement of one stakeholder in the supply chain network stops when the next stakeholder takes over. Only the produce 'knows' what happens. For the traders on the short term, this lack of transparency offers opportunities for great bargains. On the long term, this lack of transparency, due to an impaired market insight potentially erodes the mechanism of supply and demand undermining the present production system of the sector. As a result, the Netherlands vegetable market might become totally dependent on import [10, 2].

SamenMarkt® must accommodate varying lengths of engagement from all stakeholders to avoid the supply chain network becoming into disjointed intervals where unpredictability reigns and no one has authority. An effectively planned period of involvement among stakeholders enables feedback, learning, and adjustment, ultimately fostering trust.

- **Integrating rhythm**: In a supply chain network, rhythm in interaction between the different stakeholders is crucial for success. Such rhythm can be orchestrated for different reasons. The rhythm in the current horticultural fresh food supply chain network is defined by retail's demand for 'pallets' of produce [7]. Traders seek trailers, inside the Netherlands or outside. Growers have a small window within which they option to determine when to place their products on the market, when to make the products available for which price. The longer they wait in a week, the lower or higher the prices may become. Consumers are often completely unaware of how produce comes to the shelves in the supermarket. They demand good quality and divers products for little money throughout the whole year [11], where a few years ago consumers knew the season of specific vegetables and aware of the character of harvests. Also, consumer behaviour is only measured by data of buying and not by qualitative research about quality of food, ways of producing food [12]. The rising economic significance of biological food produce indicates that consumers are sensitive to these issues.

SamenMarkt® must find new methods to incorporate rhythm into the supply chain network. Novel methods of sharing information across all parties in the supply chain network can lead to innovative ways of establishing these patterns. Autonomy of each stakeholder and the interdependency of stakeholders must be supported in innovative ways.

- **Synchronizing performance**: When doing business, stakeholders have to synchronize their performance during negotiation. Currently for the growers, the traders and the retailers, such synchronization is solely determined by price strategy and politics. There is no rhythm on which they can build trust. It is a 'hit and run' climate in which each party hopes to survive. There is no synchronization with the practice of the grower. All adapt to the intensely fluctuating prices. For example, the flower auction FloraHolland, owned by the growers, provides a more stable market. Consumers do not need to synchronize their performance with that of the growers: they buy what is available. To a given extent, however, growers synchronize their performance with the consumers, for example near Mother's Day or Valentine's Day by storing products in order to have sufficient supply for the enormous demand, with a consequent deterioration of the quality.

SamenMarkt® must provide innovative methods for coordinating performance. Expanding on rhythm and aligning it with the development of products as well as the demands and preferences of consumers. Price alone should not be the sole factor influencing synchronization among partners in the supply chain. Quality and collaboration are also catalysts for establishing a sustained rhythm. *[10].*

- **Making moments to signify**: Consumers can enjoy buying the right produce, at the right time enabling them to prepare good food at home. Retailers and traders do not celebrate or mourn specific moments in relation to the products other than those related to price. For growers, the relation with the product is much deeper. In family gatherings, in church and other moments when people gather, their produce is discussed. In local communities, harvest feasts mirror important moments. Towns celebrate, for example, the first white asparagus. In a different way, supermarkets use this need for 'making moments to signify' in marketing campaigns around Christmas, Easter, the Sugar Feast, etc. connecting certain products to specific celebrations.

2.3. Relation

- **Role:** Currently, roles of all stakeholders are very distinct, but also limited to tasks at hand. Growers produce the products and are not involved in the organization of the market. Traders and retailers determine price, and most often do not directly contribute to how and which kinds of products with specific quality are grown. Consumers buy the produce offered. This division of roles seems straightforward, yet the fragmentation and the lack of healthy communication as a result cause the instable market dynamics in price and quality that now characterize the horticultural fresh food market [1].

SamenMarkt®needs to be designed for participation of all stakeholders with specific roles in the supply chain network to enable communication between different roles that supports interaction, transparency and situated agency.

- **Reputation:** Communication between growers and consumers is fundamental for building a reliable reputation system. For example, branding enables growers to communicate with consumers, which can create a reputation. Current market dynamics, however, mostly prevents such communication. Unless the grower is involved with the traders and retailers, prices of the quality products are subjected to the 'wild west' market dynamics that currently characterize the price fights between traders and retailers [13].

SamenMarkt®needs to build a reliable reputation system based on quality of fresh food products facilitating communication between growers and producers and allowing traders and retailers to contribute in constructive ways to a shared reputation system.

- **Engagement:** Currently, engagement is intense and defined by the specific task and function that stakeholders hold in the supply chain network. Frustration and lack of trust, lack of information and lack of communication, currently, define this engagement, creating an unbalanced market. The banks, not part of the analyses of this chapter, play a disruptive role in stakeholder engagement by forcing 'debt politics' into the supply chain network. The huge financial debt of some growers defines their engagement in the supply chain network. Their panic and stress undermine the market (personal communication of several bankers).

SamenMarkt®needs to create a reliable and trustworthy environment in which all stakeholders participate in which exploitation and betrayal are not rewarding. SamenMarkt®needs to build new alliances between stakeholders in the supply chain network.

- **Communion:** No processes currently exit where all stakeholders together create meaning and significance in and about the supply chain network as a whole. Growers meet at family parties, traders and retail meet at conferences and fairs and consumers see each other shop in the supermarket. In other times and in other places, celebrating a harvest was common and fundamental to the community. Such celebrations inspire the community and help people to sustain and survive the hardship of life to which the hunting, gathering and production of food largely contribute. The well-known tragedy of the commons [14] indicates the need for processes of shared meaning, also for mundane things like food. Such processes of meaning help people to know what is good to do and what is not good to do. This can effectively balance the psychological realities with the sociological and economic realities in a shared ethical framework.

SamenMarkt® must streamline the processes of establishing 'shared meaning'. Cultural knowledge assists in determining appropriate and inappropriate actions within the social and economic context of the supply chain network, facilitating shared understanding. Processes that establish common meaning reflect the ethical aspect of market dynamics. *[15]*.

2.4. Action

- **Tuning (i.e. synchronizing physical and mental actions with the requirements of the context):** When growing horticultural fresh food, one needs to tune to the growing process and one needs to be sensitive and adapt to environmental impact and other factors for doing the right thing at the right time with the growing tomato. Growers tune to their produce. Consumers, when preparing food, also 'tune' to the produce at hand. Can it be used for a salad, or is it too ripe already and should it be used for a curry? Retailers also tune to the materiality of the food. Fresh food needs a specific ripeness when sold; unsold goods need to be stored and may have to be presented differently (in combination with other goods, as prepared food already, as a bargain, etc.). Traders, however, do not tune to the materiality of the food. Food has a monetary value. For traders, scarcity and abundance are price drivers. Also, when the Dutch market does not have what they need, at the right price, traders import or export goods from the rest of the European market and beyond. The market is also changing: the demand for locally produced food, for example, is increasing significantly both in Europe as well as in the USA [16].

SamenMarkt® needs to include representation of the materiality of fresh food produce making it possible for all stakeholders in the network, including the traders, to tune to each other and to anticipate necessary adaptation.

- **Reciprocity:** (i.e. mutual dependence [17]): Reciprocity contributes to the establishment of trust only within certain circles of growers. Within these circles, growers share information, share tools and ideas with specific other growers, for example within growers' associations. Reciprocity plays a small role in the larger supply chain network. Stakeholders do business with each other, but the increasing need for sourcing reliability of the retailers and their service providers induces a trend towards more vertical reciprocity in the supply chain.

SamenMarkt® must enhance the exchange of information and communication in a reciprocal manner. This culture will require time to evolve. A culture can evolve, particularly when focusing on creating significant occasions.

- **Negotiation:** All stakeholders in the supply chain network negotiate with each other. These negotiations are characterized by hard market dynamics focused on the short-term profit, without any consideration of negative consequences for each other. Prices are defined by market dynamics in which production costs are not considered to be relevant. The price is usually set at the start of the week by the major retailer in North-West Europe (personal communication by traders). Surplus of products is either just destroyed or causes erosion of margins [7, 18]. The results of negotiations create an unbalanced market in which prices fluctuate significantly per day. As a result, growers' companies face bankruptcy and Dutch food markets are increasingly dependent on import and export. Traders and retailers maximize their profit, and consumers are mostly unaware of these dynamics.

SamenMarkt® must introduce new pricing strategies based on manufacturing and distribution expenses, as well as anticipated consumption. Price mechanisms should establish a more equitable market where all parties can achieve a satisfactory income.

- **Quality of deeds:** Different activities in the supply chain network are specific for each stakeholder and are defined by producing specific quality and quantity of the produced and traded goods. The only actions that define what happens next in the network are price negotiations. Investment in new qualities of products, innovation techniques and new alliances is currently hindered by the hard price dynamics.

SamenMarkt® must create opportunities for innovation in product quality, growth, branding, and marketing strategies. When price dynamics stabilize, a conducive environment will arise, allowing stakeholders to take calculated risks in investing in innovation.

2.5. YUTPA Analyzes Discussion

The analyses provide indicative values for each factor in the four dimensions, presented in a YUTPA graph in Figure 1. greatest trust is attained when all elements reach their greatest value. The YUTPA graph shows that trust in the present market primarily depends on players' positions in the network and their discussions. Aligning performance and quality of actions helps build trust in the present circumstances.

SamenMarkt® will need to be designed to focus on designing specific factors in each dimension for re-establishing trust in the horticultural supply chain network. Between these factors, trade-offs can be established.

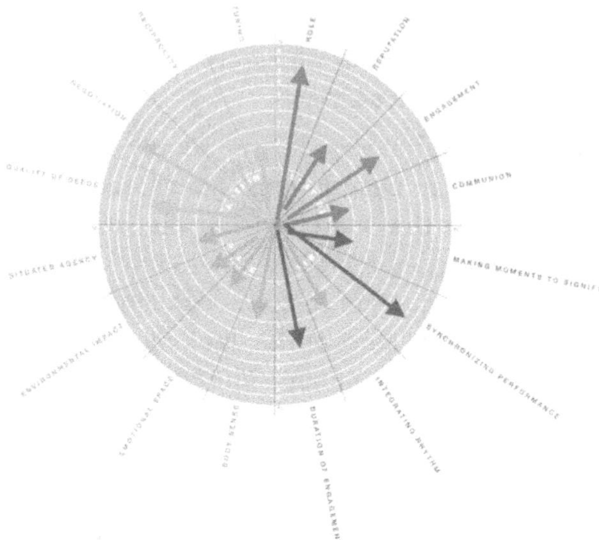

Figure 1. The YUTPA analyses of the current supply chain network in the horticultural fresh food sector show a significant lack of trust between participants (the length of the arrows are indicative for the Dutch situation and based on interviews, they are not quantitative).

In the dimension of 'relation', a better reputation system can offer more transparency in the relations between participants. As a result, also 'roles' and 'engagement' will improve and participation in the supply chain network will engender more trust. Participatory systems can offer such a reputation system. Designing the time dimension has great potential for improving the establishment of trust in the supply chain network. An elaborate design of 'integrating rhythm', 'synchronizing performance' and 'duration of engagement' can facilitate a larger awareness of being part of the larger whole of the supply chain network in which one participates. Communication in participatory systems can facilitate such trust interventions in the time dimension for all stakeholders involved.For the place dimension, an elaborate design of situated agency in the different roles and responsibilities at distinct times will generate a more balanced emotional space in which awareness of others—in other places yet part of the same supply chain network—contributes to establishing trust. A participatory system can design such situated agency and the ability to adapt situated agencies over time.

In the realm of 'activity', it is essential to ease the coordination of activities among the many actors. Enhancing tuning often leads to mutually beneficial outcomes in talks. Participatory systems can aid in adjusting by communicating and predicting the development of other participants in the supply chain network. SamenMarkt® must encourage innovation in the quality of actions and activities by providing time and space for this purpose and establishing a method to validate these efforts. Currently, 30 interviews are being conducted with stakeholders in the horticulture fresh food supply chain network to gather precise design requirements for SamenMarkt® and the transformation process.

3. PARTICIPATORY SYSTEM FOR THE FRESH PRODUCE MARKET

Participatory systems in today's networked world are defined by the capacity, magnitude, and rapidity of disseminating information and communication enabled by technology [19, 20, 21, 22]. Participatory systems are social-technical systems designed to support participation through engagement, empowerment and trust, enabling participants to act and take responsibility for their actions.

SamenMarkt® is to be a participatory system—a social technical system enabled by distributed information and coordination technology. Modelled and implemented as a multi-agent system, SamenMarkt® is a large-scale distributed market system in which software agents (small piece of code that can run on a computer as well as in the cloud and act autonomously on behalf of a user) representing their own stakeholders, negotiate terms of trade and agreement, in multiple local markets across the world. To this purpose, agents are able to reason, learn and interact with other software agents and human stakeholders. Agents reason with their own knowledge, with the knowledge they have received from other agents and with the knowledge they have observed and acquired from their interactions with the environment, and act [23]. Agents are capable of negotiating on behalf of their owners, in distributed markets at negligible cost and speed, given the relevant knowledge, and rules of the game. They can, for example, bid at auctions, and/or negotiate the terms of an agreement with a consumer or a retailer. Agents can also be designed to organize the logistics of transportation and/or financial transactions, given sufficient knowledge. Agent technology is currently being used within many existing markets in which negotiations are multiple (e.g. between producers and consumers in the energy market) and complexity high [24]. Agents are also used for negotiations in which identity is to be protected, e.g. chemical auctions, they negotiate agreements and perform transactions in supply chain networks, they regulate production lines, control production of oil and gas productions, perform scheduling for call centres, provide recommendations and plan trips.

How can a multi-agent market system for the horticultural fresh food market be designed? We are at present developing a market simulation with agent-based modelling in order to determine the prerequisites for such a market. In the last phase of our project, different market mechanisms will be designed and evaluated ranging from distributed auctions to 1-1 negotiation. Growers will be represented by their own agents—agents that have been instructed about the produce they are offering, the minimum price, quality specifications, volume, shelf-life, date of availability, sustainability, etc. Stakeholders (retailers or consumers/groups of consumers) wishing to procure produce are represented by their own agents each of which knows their owner's wishes and preferences. Logistics providers in turn have their own agents with their own knowledge of transport, price, ecology friendliness, speed and services with which they can negotiate propositions. And distribution centres, retail stores and/or online sales providers can offer their services through their own agents.

AgentScape® hosts these agents and their negotiations. Once agents have devised one or more (possible) agreements between producers, logistic providers, distribution agencies and/or consumers these agreements, commitment is needed. Commitment can either (1) automatically be effectuated by the stakeholders with prior authorization from their owners or (2) sent to the stakeholders whom, in turn, close the deals themselves with the parties involved. This second option is the option to be pursued in this project. Note that details of an agreement are only known to the participants involved.

However, knowledge can be anonymously aggregated and made available to all stakeholders, e.g. total market supply and demand, and average price at any one moment in time and over time (aggregated from all local markets), average price. This information would make it possible for all stakeholders (and their agents) in this system to be well informed with insight into the fluctuations in supply, demand and price over any given period of interest, i.e. business intelligence. This situation is comparable to stock market exchanges, where supply and demand are known at any point in time, but not the details of the individual buyers. The same is possible for the fresh vegetable and fruit market. Note that the SamenMarkt® will also need to include agents for logistics of delivery.

This is a complex market, but with the distributed internet technology, it is feasible to implement a digital trading platform that can comply with all the previous constraints, i.e. SamenMarkt®. It will create an open market situation, where price information is transparent, the transaction costs are minimal and all necessary market information is available free of charge. Provided the interface to use this agent-based market is user friendly, the financial transactions are sufficiently secured and if a sufficient number of actors take part in this market, the trustworthiness of this marketplace will create more room for a better and more innovative and sustainable (less waste) development of the entire fresh food network [25].

4. PROS AND CONS OF SAMENMARKT®

An first SWOT analysis, based on professional expertise and completed by writers, reveals the following concerns.SWOT analysis is an acronym for strengths, weaknesses, opportunities, and threats and is a structured planning method that evaluates those four elements of an organization, project or business venture

[26]. By analysing the interviews (30 in total) and scoring recurring remarks, we performed the SWOT analysis according to Panagiotou [27].

Pros:

- Supply and demand are real
- Price speculations become obsolete
- Reduction in transaction costs
- Open communication between all actors in the network
- Producers can communicate the story of the products
- Consumers can communicate explicit and latent desires
- Market intelligence is available for everyone
- Trust in the supply chain is re-established.

Cons:

- A large proportion of stakeholders need to participate for SamenMarkt® to be successful
- Transition to 'sharing in information' is difficult without visible gain beforehand
- Large-scale reselling becomes visible
- Physicality and quality of products might need to be certified
- Financial transactions need to be anonymous, while the system needs to offer transparency
- Laws on competition regulation by the Dutch and European authorities.

Opportunities:

- Tuning of production with the expected consumption
- Diminution of waste
- Efficiency of logistic services
- Incorporating physical measurements in the logistic process
- Quality controlled logistics and certification
- Optimal positioning of products for consumption
- Creating a sustainable supply chain network in which all stakeholders can survive and be well.

Threats:

- Obliterate investments in present day supply chains

- Fear of losing position in the new market place
- Neophobia, i.e. fear of new things/situations.
- Insufficient ease of use
- Not enough flexibility
- Not enough financial security.

The SWOT analysis of SamenMarkt® indicates a 'high risk-high gain' concept [27]. The business model will need to incorporate an elaborate transition plan for SamenMarkt® to be successful. Special attention is needed for the user interfaces for the different stakeholders and transparency of the overall system. The initiated advisory board in which both growers and traders are represented helps to identify possible issues in time and also helps with creating support for this research.

5. RESEARCH DESIGN FOR THE CREATION OF SAMENMARKT®

The following steps have been identified in the iterative design process for SamenMarkt®.

Phase 1

- Formulation of concept SamenMarkt® on the basis of YUTPA analyses of the current supply chain network of tomatoes.
- Funding and partners are identified and approached, the advisory board is created. The advisory board composed of industry experts functions as sounding board for the research.

Phase 2

- Thirty interviews identifying local dynamics in the supply chain have been carried out to validate the initial YUTPA analysis and fine tune design requirements (Figure 2).
- The first simulation model is created. Advisory board of horticulture sector guides this simulation.
- Validation of the first simulation in workshop with different stakeholders in the supply chain network of the horticulture sector in the Netherlands.

Phase 3

- Development business case and governance structure of SamenMarkt®.
- Second distributed model by design team and advisory board of the horticulture sector.
- Series of validations with different stakeholders in horticultural supply chain network of fresh food.

- Orchestration of transformation processes.

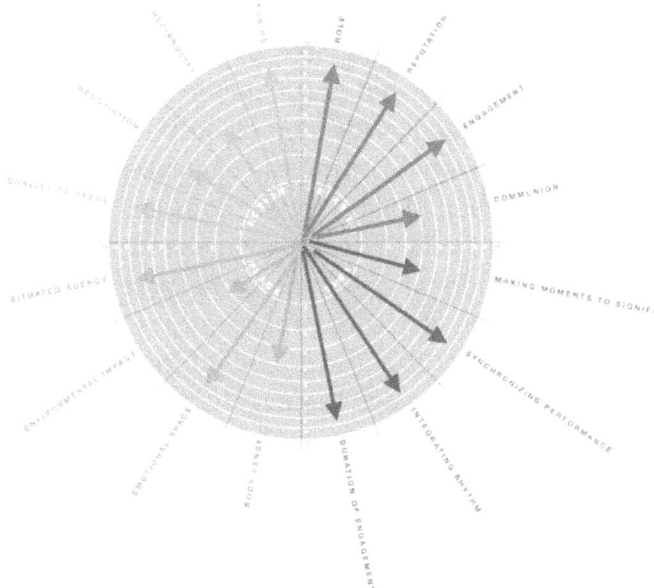

Figure 2. Using the YUTPA framework to identify design requirements for SamenMarkt®, a new solution space for the horticultural fresh food sector.

Phase 4

- Identifying investors, owners and shareholders SamenMarkt®
- Acquiring support form societal and business stakeholders (government, financial sector, international partners)
- Establishing business and governance structures
- Introduction orchestration transition to SamenMarkt®.

TU Delft and HogeschoolInholland conducted and funded the initial phase of SamenMarkt® from 2012 to 2014. The second phase commenced in January 2015 and concluded in June 2015. During the second phase, additional money and alliances are obtained. Phase three is anticipated to span from September 2015 to September 2017, following which phase four can commence. By September 2020, SamenMarkt® aims to become a sustainable market where all participants can thrive and be prosperous.

6. CONCISE SUMMARY

The current simulation model of the tomato market in North-West Europe, created during phases two and three of this study, demonstrates that distrust and self-interest lead to significant losses and substantial price swings throughout the supply chain. The chain stifles innovation at the first stages. Most vegetable commodities are experiencing a downward trend in profit margins. A digital trade platform, even in its basic form, can positively impact price stability by reducing emotional elements that currently affect trading. An efficient distributed digital trading platform adopted by most participants in the supply chain will enhance productivity, minimize duplication, and boost confidence in the vegetable industry, leading to a more sustainable food supply chain. Future study will concentrate on adapting the SamenMarkt® concept to broader and more intricate international food markets. Additional research will determine if the notion of SamenMarkt® may indeed play a significant part in the growing global food issues. The authors of this chapter propose that SamenMarkt® will facilitate the reestablishment of trusted relationships between producers and consumers. This platform will utilize distributed information and communication technology (ICT) to efficiently manage the negotiation of prices and distribution of products by handling large, complex real-time data through accessible and effective interfaces.

7. AT A GLANCE

A crisis is developing in the horticulture fresh food supply chain network in the Netherlands. During the period of 2011–2013, 50% of growers were unable to meet their financial obligations, leading to an imbalance in the market and potential bankruptcies. Trust among participants in the supply chain network has diminished. This chapter introduces the concept of the not yet established and outlines design requirements for new systems to tackle this difficulty and offer guidance for potential enhancements. This chapter presents the concept of SamenMarkt®, a participatory system that utilizes multi-agent system technology to provide distributed price negotiation, distribution, and communication among producers, merchants, and consumers. An study of the strengths, weaknesses, opportunities, and threats of the SamenMarkt® concept is presented along with a research and development strategy that utilizes simulation and emulation to engage stakeholders and participants in designing a decentralized digital marketplace. The upcoming research will investigate how SamenMarkt® can offer a solution for the growing global food issues. We are currently utilizing agent-based modeling to replicate the current market and scenarios. The next phase is to construct the agent-based platform for real-time negotiations and business intelligence.

REFERENCES

1. Verheul J. Kans voorkrachtigerafzetorganisaties. Groenten& Fruit: Glasgroenten. McKinsey. 2014;1499:1
2. van Velden P. De NederlandseTuinbouw - End of an Era II. Foodlog.nl 30 August 2014;14:42. Available from: https://www.foodlog.nl/artikel/the-end-of-an-era-ii/allcomments/

3. Nevejan C. Witnessed presence and the YUTPA framework. Psychnology Journal. 2009;7(1):59-76

4. Nevejan C, Brazier F. Granularity in reciprocity. AI & Society. 2012;27(1):129-147

5. In a workshop between professor Olaf van Kooten en dr. Caroline Nevejan in 2012, as result of a YUTPA analyses of current dynamics in the horticultural supply chain in the Netherlands, the concept of SamenMarkt® emerged. Together with professor Frances Brazier, drs. Coen Hubers and dr. Michel Oey the concept of SamenMarkt® has been further elaborated

6. Ondersteijn CJ, Wijnands JH, Huirne RB, van Kooten O, editors. Quantifying the Agri-food Supply Chain. Springer, Dordrecht, The Netherlands; 2006

7. van Uffelen RL, Splinter GM, van Kooten O, Botden N. Waste reduction in the supply chain of arable vegetables. In: XV International Symposium on Horticultural Economics and Management Acta Horticulturae 655, 283-289 DOI: 10.17660/ActaHortic.2004.655.35

8. Schepers H, van Kooten O. Profitability of 'ready-to-eat' strategies". Towards model-assisted negotiation in a fresh-produce chain. Agrotechnology and Food Innovations, Wageningen University and Research, Springer Verlaag, Dordrecht, The Netherlands; 2006

9. Yu M, Nagurney A. Competitive food supply chain networks with application to fresh produce. European Journal of Operational Research. 2013;224(2):273-282

10 Van Kooten O, Kuiper E. Consumer acceptability in flower chains: How can we determine what the final customers really want? In: IX International Symposium on Postharvest Quality of Ornamental Plants Acta Horticulturae 847, 17-26 DOI: 10.17660/ActaHortic.2009.847.1

11. Mitchell VW. Consumer perceived risk: Conceptualisations and models. European Journal of Marketing. 1999;33(1/2):163-195

12. van der Vorst JG, Schouten RE, Luning PA, van Kooten O. Designing new supply chain networks: Tomato and mango case studies. In: Horticulture: Plants for People and Places. Vol. 1. The Netherlands: Springer; 2014. pp. 485-501

13. Wilson N. The supply chains of perishable products in northern Europe. British Food Journal. 1996;98(6):9-15

14. Harding G. Tragedy of the commons. Science. 1968;162(3859):1243-1248

15. The world has enough for everyone's need, but not enough for everyone's greed. Quote of Mahatma Ghandi

16. Kramer MR. Creating shared value. Harvard Business Review. 2011;89(1/2):62-77

17. Nevejan C, Beazier F. Time design for building trust in communities of systems and people. In: ICoRD '11 International Conference on Research into Design; 10-12 January 2011; Indian Institute of Science Bangalore

18. Conner DS, Montri AD, Montri DN, Hamm MW. Consumer demand for local produce at extended season farmers' markets: Guiding farmer marketing strategies. Renewable Agriculture and Food Systems. 2009;24(04):251-259

19. Ogston E, Brazier F. Agentscope: Multi-agent systems development in focus. In: The 10th International Conference on Autonomous Agents and Multiagent Systems. Volume 1. 2011 May 2. pp. 389-396. International Foundation for Autonomous Agents and Multiagent Systems

20. Rezaee SA, Oey M, Nevejan C, Brazier F. Participatory demand-supply systems. Procedia Computer Science. 2015;44:105-114

21. Castells M. The Rise of the Network Society: The Information Age: Economy, Society, and Culture. Cichester, Sussex, United Kingdom. John Wiley & Sons; 2011

22. Thackara J. In the Bubble: Design in a Complex World. Camebridge, Massachusetts, London England, MIT Press; 2005

23. Pournaras E, Warnier M, Brazier F. A distributed agent-based approach to stabilization of global resource utilization. In: Complex, Intelligent and Software Intensive Systems, 2009. CISIS'09. International Conference on 2009 March 16; Los Alamitos, CA USA. pp. 185-192

24. Brazier FM, Cornelissen F, Gustavsson R, Jonker CM, Lindeberg O, Polak B, Treur J. A multi-agent system performing one-to-many negotiation for load balancing of electricity use. Electronic Commerce Research and Applications. 2002;1(2):208-224

25. Brazier F, Nevejan C. Vision for Participatory Systems Design. Procedia Computer Science, 2015;44:105-114, ISSN 1877-0509; 2014

26. Pickton DW, Wright S. What's swot in strategic analysis? Strategic Change. 1998;7(2):101-109

27. Panagiotou G. Bringing SWOT into focus. Business Strategy Review. 2003;14(2):8-10

03 CITRUS VALUE CHAIN(S): SURVEY OF THE CITRUS INDUSTRY IN PAKISTAN

1. PAKISTAN CITRUS INDUSTRY

Agriculture is crucial to Pakistan's economy and is essential for success. Pakistan's agricultural sector benefits from abundant agricultural resources, fertile land, well-irrigated plains, and diverse seasons. While the agriculture sector's GDP contribution has declined, almost two-thirds of the population still depend on it for their living [1]. Agriculture is acknowledged as a significant factor in the country's economic success. The total agricultural crop production in 2014–2015 was expected to reach 116 million tons. Pakistan annually produces around 13.5 million tonnes of fruits and vegetables. In 2014–2015, the total fruit production was 7.01 million tons, accounting for 48.3% of the country's total fruit and vegetable production[2, 3].

1.1. Citrus production

Fruit production in Pakistan has been increasing overall, except for a notable decline in 2006-2007. The decrease was linked to unfavorable weather circumstances, such as a hailstorm and lack of water. Both the acreage dedicated to fruit cultivation and fruit output have been consistently increasing. Citrus fruit has the highest output volume, followed by mangoes, dates, and guava. In 2014-2015, citrus production reached 2.4 million tonnes, representing 35.2% of total fruit production[3]. Citrus fruits such as mandarins (Kinnow), oranges, grapefruit, lemons, and limes are important in Pakistan, with mandarins (Kinnow) holding particular significance.

Pakistan's total production of citrus fruit (primarily Kinnow) is approximately 2.0 million metric tonnes annually. Although there is no remarkable increase in area under citrus production, the production has increased up to 30.8% since 1991–1992. In 1991–1992, Pakistan produced 1.62 million tonnes citrus, which increased to 2.1 million tonnes in 2008–2009 and 2.4 million tonnes in 2014–2015 [3].

Citrus fruit production increased from 1993-1994 but started to decline in 1999. The citrus fruit crop requires a precise low temperature for healthy development. Not achieving this temperature can lead to decreased fruit production. [4].

Temperature variations in the citrus-growing areas of Pakistan could be a contributing factor to the inconsistent citrus fruit yield. During 2006-2007, there was a notable temperature change in citrus-producing areas, resulting in a fall in citrus yield from 2.4 to 1.4 million tons. However, the size of citrus fruit orchards did not change[3].

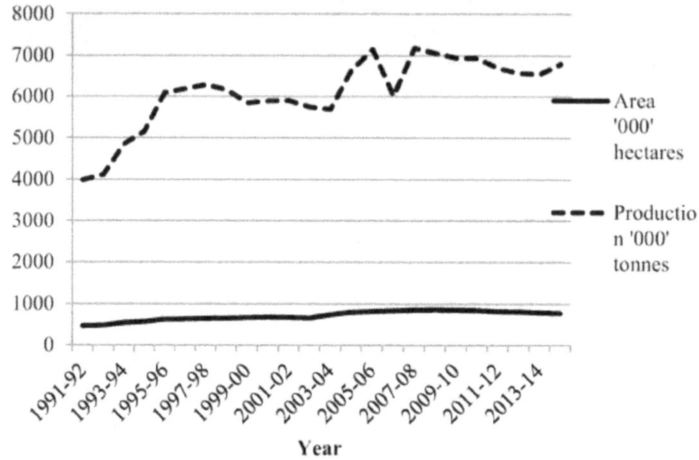

Figure 1. Area and production of all fruit in Pakistan. Source: [3, 5, 6].

Citrus fruit is mostly grown in four provinces of Pakistan: Punjab, Khyber Pakhtunkhwa (KPK), Sindh, and Baluchistan. Punjab is acknowledged as the main hub for citrus cultivation among the four provinces. Table 1 shows the main citrus growing areas in all four provinces of the country.

Punjab province is responsible for over 90% of the overall Kinnow production in Pakistan, as stated by the Pakistan Horticulture Development and Export Company (PHDEC) in 2005. In contrast, KPK mostly produces oranges among all citrus fruits in the country. Sargodha, Toba Tek Singh, and Mandi Bahauddin are renowned areas in the Punjab province for their citrus cultivation. Citrus fruits of various types are cultivated in small quantities in other districts as well.

Punjab province, according to Pakistan Horticulture Development and Export Company (PHDEC) (2005), produces more than 90% of total Kinnow production whereas KPK mainly produces oranges among all citrus fruits in the country. Sargodha, Toba Tek Singh and Mandi Bahauddin are three known districts for their citrus production in Punjab province. Different varieties of citrus fruits are also grown in small proportions in other districts.

Mandarins (Feutrell's Early and Kinnow) and sweet orange (Mausami or Musambi and Red Blood) are very important among all the citrus varieties cultivated in Pakistan. Table 2 shows different varieties of citrus produced in the country. Punjab province, being the hub of citrus production (Kinnow), produced 97.1% of citrus fruit (Kinnow) in 2014–2015 [3].

In Punjab province, three districts, Sargodha, Toba Tek Singh and Mandi Bahauddin, constitute around 55% of the total area under citrus cultivation and produce nearly 62% of citrus fruit [6]. In Sargodha district, Bhalwal produces 650,000 metric tonnes of Kinnow annually and is considered as the centre of Kinnow (mandarin) production (Pakistan Horticulture Development & Export Company (PHDEC), 2005). Figure 2 demonstrates citrus fruit production in four provinces of Pakistan.

Table 1. Major citrus growing areas in Pakistan.

Province	Major districts
Punjab	Sargodha, Toba Tek Singh, Mandi Bahauddin, Sahiwal, Khanewal, Vehari, Bahawalpur, Multan, Okara, Layyah, Jhang, Kasur, Bahawalnagar, Faisalabad
Khyber Pakhtunkhwa (KPK)	Malakand, Swat, Nowshera, Lower Dir, Dera Ismail Khan, Mardan, Haripur
Sindh	Naushero Feroze, Khairpur, Nawabshah, Sukkur, Sanghar
Baluchistan	Nasirabad, Dolan, Lasbela, Gwadar, Sibi,
Source: [7, 8]	

Table 2. Varieties of citrus fruit in Pakistan.

Sweet Orange	**Succri, Musambi, Washington Navel, Jaffa, Red Blood, Ruby Red and Valencia Late.**
Mandarins	Feutrell's Early and Kinnow
Grapefruit	Mash Seedless, Duncan, Foster and Shamber
Lemon	Eureka, Lisbon Lemon and rough Lemon
Lime	Kagzi lime and Sweet lime
Source: [8]	

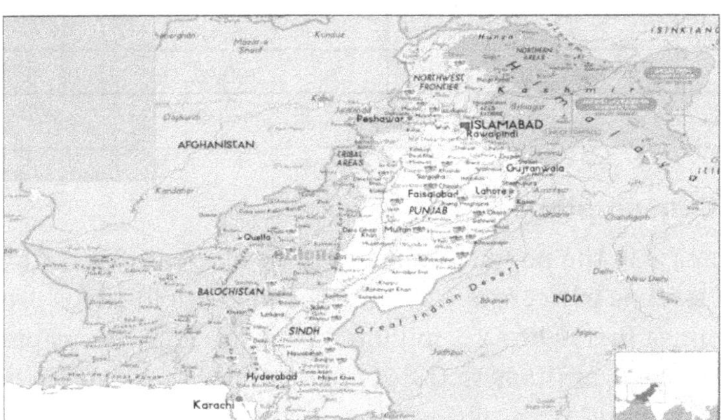

Figure 2. Province-wise production of citrus fruit in Pakistan. Source: [3, 6].

In 2014–2015, the total citrus production in Khyber Pakhtunkhwa (KPK) was 1.29%, in Sindh 1.26% and in Baluchistan it was 0.29%. In the late 1990s, the production of citrus fruit in the Baluchistan province increased and it was due to the increase in the area under citrus cultivation and hence production under citrus fruit was increased in Baluchistan in the late 1990s [3, 6, 9].

In Punjab province, Kinnow production was 1.80 million tonnes followed by oranges, which was about 94 thousand tonnes in 2009–2010 among different varieties of citrus fruits. In Punjab, Kinnow was 87.1% of the total citrus production and 80.3% of the total area under citrus cultivation. Oranges come next to Kinnow in production and area under cultivation and constitutes 4.5% of the total citrus production and 6% of the total area under citrus cultivation in Punjab. Grapefruit production is the lowest and it was only 3 thousand tonnes. Different types of citrus grown in Punjab province in 2009–2010 are shown in Table 3.

Table 3. Production of different types of citrus fruit in Punjab.

Type of Citrus	Production and area of different types of citrus fruit in Punjab 2011–2012		
	Area ('000' Hectares)	Production ('000' tonnes)	Percent of total citrus production
Kinnow	154.6	1876.0	89.43
Oranges	9.4	80.0	3.81
Musambi	7.2	61.5	2.93
Mandarin	1.2	9.3	0.44
Sweet Lime	3.9	29.5	1.41
Sour Orange	0.1	1.0	0.05
Lemon	4.6	26.0	1.2
Sour Lime	0.8	3.8	0.2
Grapefruit	0.3	2.3	0.1
Other	1.2	8.4	0.4
Total	183.2	2097.7	100.00
Source: [10]			

1.2. Citrus Consumption

The rapid increase in the population may be one of the major factors keeping the consumption level nearly constant despite the increase in citrus production from 1.8 million tonnes in 1999 to 2.1 million tonnes in 2009. Per capita income has increased from US$450 in 1999 to US$917 in 2009 [11]. However, the sharp decline in per capita consumption in 2007 was a result of lowest production of 1.4 million tonnes from 2.4 million tonnes in 2006, which resulted in lower domestic supply and availability of citrus fruit in the country. The high peak of 2005 in Figure 3 reflects the highest per capita consumption of citrus fruit due to decreased exports of the citrus fruit from 151.3 thousand tonnes in 2004 to 79.2 thousand tonnes in 2005.

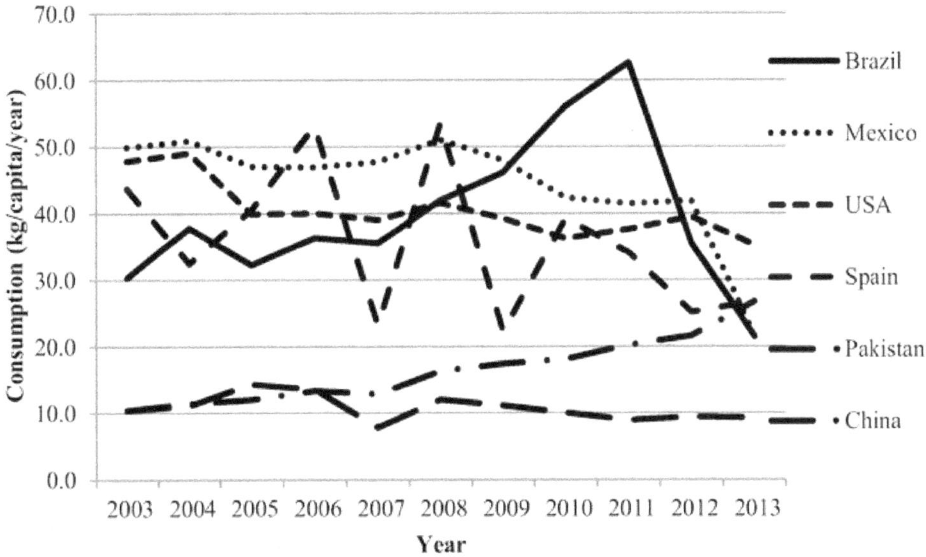

Figure 3. World citrus fruit consumption trend. Source: [12].

Figure 3 shows the annual per capita consumption (kg) of fresh citrus fruit in important countries. There was an increase in per capita consumption in Brazil that reached to the highest level in 2011 and then dropped very sharply in 2012 and 2013.

1.3. Citrus Exports

With the changing consumer preferences towards consumption of fresh and convenience food, the global demand for fresh fruit is increasing [8, 13]. Pakistan is one of the largest citrus producing countries and ranked 13th in the production of citrus fruit [5]. It has been observed that fresh citrus exports from Pakistan have been increasing since 1995–1996 as shown in Figure 4.

Kinnow mandarin accounts for around 97% of the entire citrus fruit exports from the country[8, 14].. In 2014–2015, Pakistan exported 393 thousand tons of citrus fruit valued at $204 million, accounting for around 16.4% of the total citrus production. In 2009–2010, citrus exports generated $99.4 million in foreign earnings, which was three times higher than in 2000–2001. Although output has risen, only a modest percentage (8–12%) of citrus fruit is being exported.

The majority of the farmers in Pakistan own and cultivate a small size of agricultural land of less than 2 hectare. However, in Punjab, average citrus farm size was 12.3 hectare, which is considered relatively large compared to other crops [15]. The size of citrus orchard ranges from less than 1 hectare to as big as 65 hectare in different regions of the country. A few large citrus growers do exist; however, small and medium size growers are the majority [16].

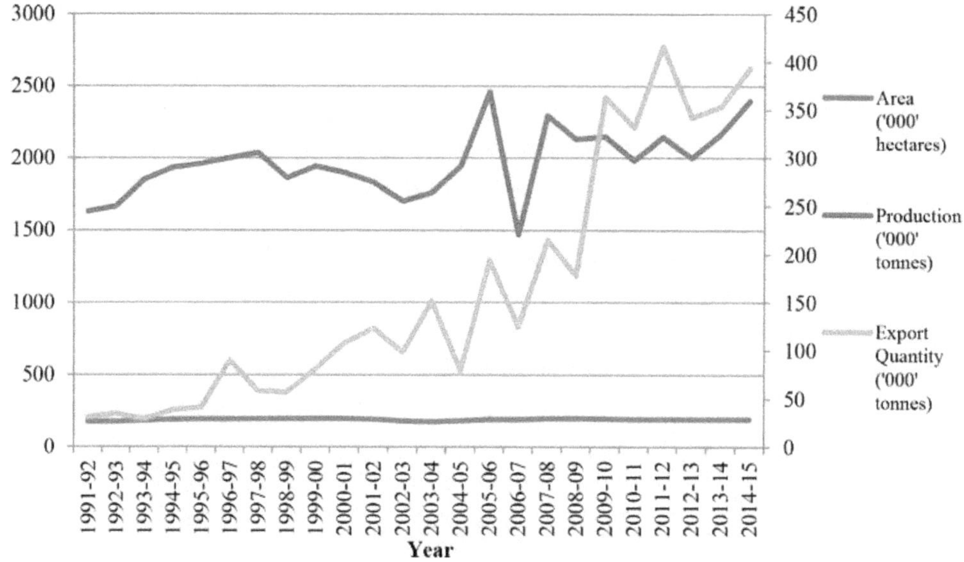

Figure 4. Pakistan citrus area, production and exports. Source: [6, 14].

Pakistan's agricultural value chains are highly varied and originate with citrus growers/producers. The citrus fruit value chain, like many other fruits, is predominantly managed by the private sector. The government facilitates commercial transactions by providing essential infrastructure and regulatory measures. Marketing intermediaries often abuse agricultural crop producers by imposing exorbitant margins on their investments. [16]. Citrus fruit value chain(s) starts with the involvement of pre-harvest contractor directly with citrus growers. According to Chaudry [15], the typical citrus value chains in Pakistan are shown in Figure 5. The role of each player in the citrus value chain is presented later in the results and discussion sections.

All stakeholders in these value chains perform their typical roles as in other food value chains, however the pre-harvest contractors hold the most significant influence in Pakistan's citrus value networks. [15, 16].

Considering the surge in production, steady domestic demand, and surplus citrus fruit available for export, the question is: How many distinct citrus value chains are active in the country? Who are the participants in the citrus value chains? Why is there a limited export of citrus fruit despite its abundance? This study intends to identify and analyze various value chains of Kinnow citrus fruit in Pakistan and examine the roles and responsibilities of each participant in the citrus sector. This report also highlights the next prospects and obstacles facing the citrus industry in Pakistan.

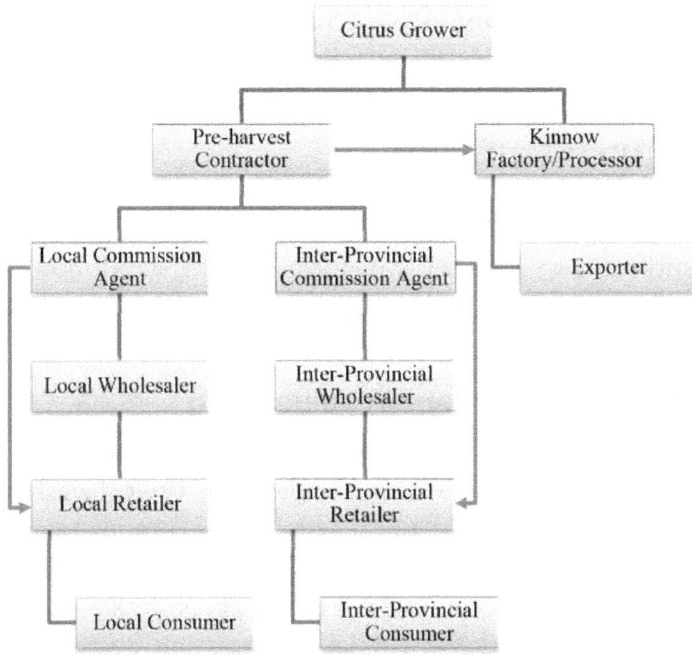

Figure 5. Value chains of citrus in Pakistan. Source: [15].

2. METHOD

The study utilized both qualitative (exploratory) and quantitative (descriptive and inferential) research methodologies based on the research topics. In 2013-2014, a survey of various participants in Pakistan's citrus business was carried out using semi-structured interviews with the aid of a questionnaire. Surveys were used to collect primary data, while secondary data were gathered from published documents, reports, journals, and government publications from different public and commercial entities and departments.Using a convenience sampling technique, a total of 245 respondents were interviewed during a period of 4–5 months from three leading citrus producing districts, namely Sargodha, Toba Tek Singh and Mandi Bahauddin. The respondents included were 126 citrus growers, 99 pre-harvest contractors and 20 processing factories/exporters from these three districts. Due to unavailability of population size (sampling frame), time and budget constraint, a convenient sampling technique was used for the selection of the respondents. The underlying reason of interviewing different players in the citrus value chains was to study closely their functions and participation in different existing value chains. It also helped identifying and understanding different value chains operating in the country. This study also identified future opportunities and challenges of citrus sector in Pakistan.

A software program designed for statistical analysis The acquired data was analyzed using Predictive Analytics Software–version 21 (PASW-21), formerly known as Statistical Product and Service Solutions (SPSS). Fisher's exact test was utilized to assess the importance of several demographic variables in relation to the selection of the value chain, specifically including the pre-harvest contractor.

3. RESULTS AND DISCUSSION

3.1. Citrus (Kinnow) Value Chain Systems

The study identified two primary categories of citrus (Kinnow) value chains: unprocessed and processed. The following section will offer an in-depth explanation of various value chains.

3.1.1. Unprocessed Citrus (Kinnow) Value Chain

Citrus fruits (Kinnow) that have not been processed are sorted and packaged prior to being sold in the local market, without undergoing typical washing and waxing procedures. Figure 6 presents the unprocessed citrus (Kinnow) value chain in the country. About 50–60% Kinnow is marketed in the country for domestic consumption while nearly 30% Kinnow is accounted for post-harvest losses [8]. Citrus growers, pre-harvest contractors, local (provincial) commission agents, inter-provincial commission agents, local wholesalers, inter-provincial wholesalers, local and inter-provincial retailers are different actors of unprocessed citrus (Kinnow) value chain in Pakistan.

Citrus growers and pre-harvest contractors play a crucial role in the Kinnow value chain, and their marketing strategies greatly influence the citrus supply chain in the country. Historically, almost 90% of citrus growers frequently sold their plantations to pre-harvest contractors. The primary causes include financial constraints, limited market data, transaction ease, hedging against price fluctuations, and corporate policies. [15]. Citrus growers are shifting towards a more market-oriented approach and utilizing various marketing channels to secure higher prices for their products, rather than exclusively selling to pre-harvest contractors. Enhanced education, governmental backing, and technology advancements are the factors contributing to this transformation. When questioned about directly marketing his fruit in the market, one of the citrus growers responded.

Now, almost every citrus grower may obtain prices from several local and provincial markets, enabling us to engage in direct marketing of our product. Due to our lack of knowledge about various market prices in the past, we were unable to determine where to sell, so we proceeded with the pre-harvest contractor. We appreciate the government's backing, which allowed us to sell our products directly and make a substantial profit.

3.1.2. Processed Citrus (Kinnow) Value Chain

Kinnow processing is of two types: (1) for export and involves washing, waxing, grading and packing and (2) for juice extraction.

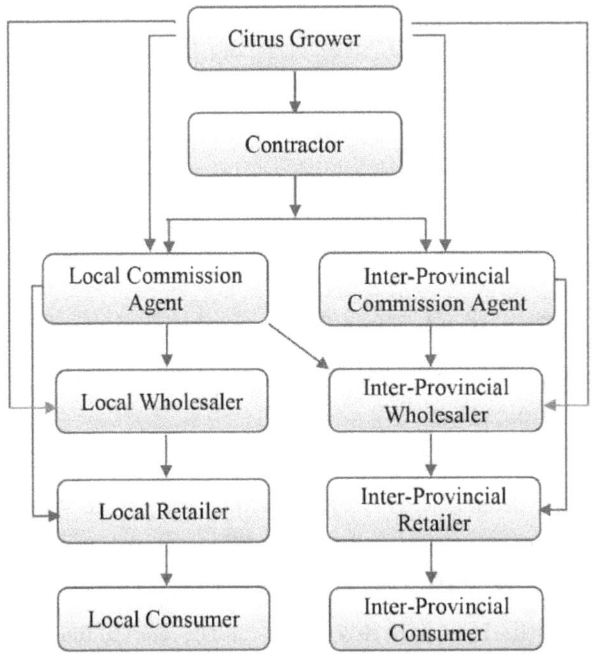

Figure 6. Unprocessed citrus (Kinnow) value chain for local market.

3.1.2.1. Processed Citrus (Kinnow) Value Chain for Export

Only 8–12% of Kinnow is processed for export to various nations. In 2013–2014, the primary exports of Kinnow mandarin were to Afghanistan (113,000 tonnes), Russian Federation (72,000 tons), and United Arab Emirates (68,000 tonnes). A comprehensive list of countries can be found in Appendix A [17]. The citrus (Kinnow) value chain includes citrus producers, pre-harvest contractors, local commission agents/wholesalers, Kinnow processing companies, and exporters. (Figure 7). The processed citrus value chain closely resembles the unprocessed citrus value chain, with the main difference being the inclusion of a Kinnow processing factory and exporters in the processed value chain during the final step of the process.

Citrus growers, pre-harvest contractors, and commission agents directly sell citrus fruit to firms involved in Kinnow processing and exporters. Pre-harvest contractors and certain commission agents represent Kinnow processors and exporters by buying citrus fruit on their behalf. These agents receive credit or advance payments to purchase citrus fruit. Kinnow processors and exporters favor vendors with a strong familial and financial history, an excellent market reputation, and the ability to consistently deliver fruit during the season. Sellers receive payment via checks from processors and exporters.

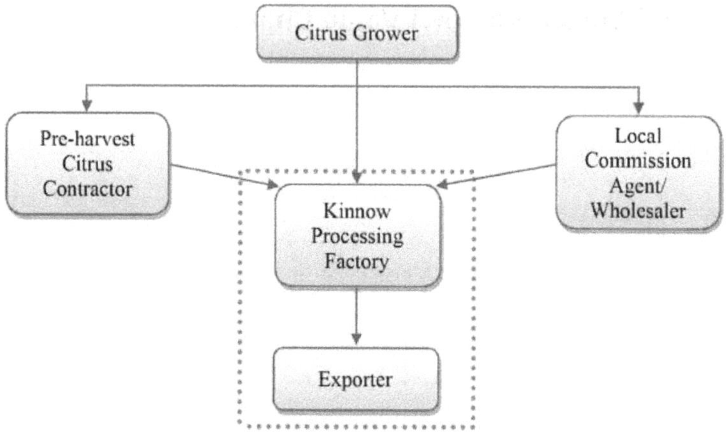

Figure 7. Processed citrus (Kinnow) value chain for export.

One of the exporters commented on the selection of seller (grower/contractor/commission agent/wholesaler):

'In our export business, we need to make promises with our importers for certain quantity of fruit in the season; therefore, all exporters try to deal with the seller who can commit constant supply of citrus. A well reputed, financially strong with good moral character seller is our first choice to deal with'.

In Sargodha, about 150 Kinnow processing factories are functioning but only 25–30 processing factories are exporting citrus to different countries. The reasons of such a low export from the country include high demand for seedless Kinnow (hybrid mandarin) in the developed countries (not produced in Pakistan), marketing practices and failure to meet the quality certification standards. One of the respondents who wanted to export to European countries commented:

'I always take good care of my orchards; hence, I manage to get good quality fruit in each season. Realising the quality, I decided to export my fruit to Europe last year but I was told that I should grow citrus to meet quality certification standards required for export which I am following now. Hopefully, I would be able to export my citrus to Europe soon'.

A few processing units are also being constructed in other districts as well, for example, district Mandi Bahauddin, district Toba Tek Singh and district Multan.

3.1.2.2. Processed Citrus (Kinnow) Value Chain for Juice Extraction

Pakistan has a total of 52 fruit processing and juice extraction facilities in operation. Only five of them are extracting juice from citrus fruit[18]. Approximately 6% of the total citrus produced in the country is utilized for juice extraction. The factories are clustered in the primary citrus-producing regions such Sargodha, Bhalwal, and Mateela cities. Citrus producers and pre-harvest contractors sell substandard fruit to juice factories, who then convert it into juice concentrate and distribute it to retail shops and supermarkets.

One of the contractors who supply to exporters and processors replied:

'I always try to sell exporters and processors because they pay premium price for my fruit. However, these exporters and processors only purchase selective fruit from me and leave small size, non-uniform and unripe fruit. I don't mind it; I sell this non-selective fruit to juice factories and recover my cost. Job done'.

Figure 8 shows different functionaries of processed citrus (Kinnow) value chain.

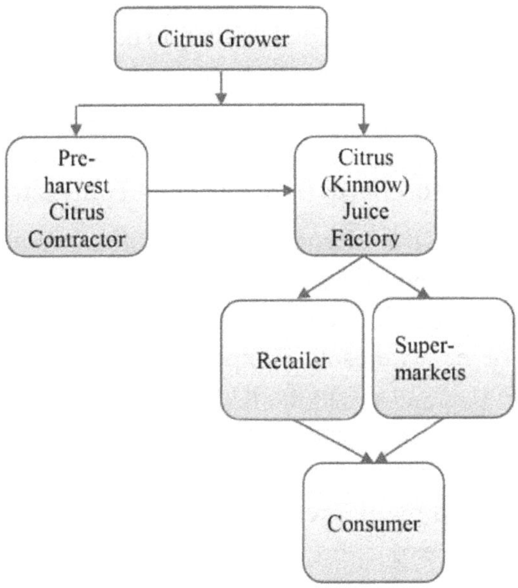

Figure 8. Processed citrus (Kinnow) value chain for juice extraction.

3.2. Description of Different Value Chain Players

Different players of citrus (Kinnow) value chain including citrus growers, pre-harvest contractors, commission agents, wholesalers and retailers are described and discussed below.

3.2.1. Citrus Growers

The results of this study revealed that majority of the citrus growers (76.2%) were 31–60 years of age and only 6.3% were quite young (under 25 years old) in the citrus industry. However, about 17% citrus growers were above 60 years of age. The literacy rate was more than 90% in the citrus growers (having at least 5 years of schooling) and only 3.2% respondents were illiterate. It was reported that the literacy rate was increasing in the citrus growing areas in Pakistan, thus providing basis for the growth of citrus growers and making them more market oriented [19]. The results also showed that nearly 80% of the citrus growers were very experienced and were involved in the citrus industry for many years. The size of the citrus orchards was small, and majority of the citrus growers possessed less than 20 acres (Table 4).

The citrus groves' size decreased as they were shared among the subsequent generations. Previously, it was discovered that over 90% of citrus growers favored selling their produce through pre-harvest contractors, which decreased their direct participation in the value chains. The literacy rate has increased in the past decade due to improved government initiatives in education and growing market awareness. All citrus growers were taught and informed about market information and opportunities. Citrus growers are increasingly seeking market information to explore alternate value chains instead of selling directly to pre-harvest contractors. Approximately 46% of citrus growers are transferring ownership of their orchards to pre-harvest contractors, as indicated by this survey. This finding is contrary to the previous study conducted by Chaudry [15] and Ali [16] who reported that 95% of citrus growers sold their orchards to pre-harvest contractors about 15 years ago.

Approximately 30% of the citrus farmers in the study sold their orchards to commission agents, wholesalers, or exporters. The pricing was determined based on a commission rate of 7–8% of the total sale price of the produce with the commission agent or wholesaler. Exporters and processing facilities often disclose the purchase price of citrus fruit per 40 kg for the season. Negotiability may be possible in specific instances, such as when citrus growers agree to a full-season fruit supply contract. Kinnow prices are not regulated by the government and are instead influenced by market dynamics such as demand and supply. The challenge when selling to exporters was that they only bought high-quality fruit that met specific criteria regarding size and color (selective fruit procurement). Citrus growers switch to alternative purchasers who acquire the entire supply at the set price due to this reason. Approximately 6% of citrus growers distributed their harvest among multiple purchasers rather than selling to a single buyer.

Table 4. Descriptive statistics for citrus growers.

Age (years)	**1.0–30**	**31–60**	**61–90**	
Percent (%)	6.3	76.2	17.5	
Education	**Illiterate**	**Undergraduate**	**Graduate**	**Postgraduate**
Percent (%)	3.2	74.6	15.1	7.1
Experience (years)	**≤10**	**11–25**	**>26**	
Percent (%)	21.4	40.5	38.1	
Area under citrus (acres)	**1.0–20**	**20.1–60**	**60.1–160**	
Percent (%)	53.2	30.2	16.7	
Sell orchard to	**Pre-harvest contractor**	**Commission agent/ wholesaler**	**Processing factory/ exporter**	**Different buyers**
Percent (%)	46%	30%	18%	6%

Table 5. Citrus pre-harvest contractors general information.

Size of contractors	Less than $0.1 million	$0.1–$0.5 million	More than $0.5 million
Percent (%)	20%	70%	10%
Experience	**Less than 10 years**	**10–20 years**	**More than 20 years**
Percent (%)	20%	45%	35%
Sell fruit to	**Commission agent/ wholesaler**	**Exporter**	**Different buyers**
Percent (%)	50%	40%	10%
Education status	**Illiterate**	**Educated to minimum level**	
Percent (%)	35%	65%	

3.2.2. Citrus Pre-Harvest Contractors

99 pre-harvest contractors were interviewed for the study. Around 70% of pre-harvest contractors had a business volume ranging from US$0.1 million to $0.5 million and were classified as medium-sized contractors. Pre-harvest contractors either purchased the orchards using their own funds or served as representatives for various stakeholders in the citrus supply chain, such as commission agents, wholesalers, processors, or exporters. Most pre-harvest contractors (80% of respondents) had over 10 years of experience, as indicated in Table 5. Small pre-harvest contractors with fewer than 10 years of experience work sporadically in the citrus market and are not regular participants in the country's supply chain. The majority of citrus growers were educated and experienced businessmen, whereas only 35% of pre-harvest contractors were illiterate, having less than two years of schooling. However, they collaborate with well-educated business partners or family members like brothers or sons. Approximately half of the pre-harvest contractors sold their fruit to commission agents or wholesalers, as indicated in Table 5. The price is paid on a commission basis at a rate ranging from 7% to 8% of the total sale price of the produce. Typically, pre-harvest contractors and commission agents or wholesalers would enter into a one-year contract, either written or verbal, to supply a specific quantity of fruit at a fixed price. Approximately 40% of the pre-harvest contractors immediately sold their produce to exporters. Typically, the price was determined by processors and exporter associations, although occasionally it was open to negotiation. Citrus producers sold their groves to contractors in advance for a variety of reasons. Citrus growers in Pakistan did not have the necessary financial means to efficiently market their produce. Farmers and pre-harvest contractors share the responsibility of managing the orchard. The contractors now offer fertilizers and herbicides, alleviating the financial burden on farmers. The pre-harvest contractor

held control over the produce and the ability to negotiate the sale of the fruit at the market. A citrus farmer discussed his decision to sell his produce prior to the harvest.

'It is always a problem for me to buy inputs once my orchard starts flowering. Pre-harvest contractor, on one hand, purchase my orchard well in time and on the other hand, help in providing me the required inputs which otherwise I cannot afford'.

3.2.2.1. Types of Contractors

Various citrus orchard contractors are active in the country. They can be categorized into three groups based on their ability to buy citrus fruit for trading.

- **Small-sized contractors**

Small-sized contractors with buying power less than US$0.1 million (1 US dollar = 85 PKR) are called '*Den Daar*' and usually work domestically in the local markets only. Being a small-scale operator and limited finances, they work in groups and buy an orchard on shared basis. They sell fruit on daily basis in the local market where they are called '*Phariwala*' in the local language or sometimes they sell directly to a wholesaler or commission agent.

- **Medium-sized contractors**

Medium-sized contractors buy fruit that value between US$0.1 to US$0.5 million. They work for commission agents/wholesalers/exporters and usually do not invest their own money. Sometimes they buy the orchard with their own money but usually they contract with the citrus grower at the pre-harvest stage by estimating the future production from the orchard and also make a future contract with the commission agents/wholesalers/exporters. This way they play an intermediary role between two parties without investing their own finances. Commission agents/wholesalers/exporters are then responsible to pay all the money to the contractor in the form of partial payments. Initially, commission agents/wholesalers/exporters pay one-fourth of the total orchard value under the contract to the contractor who pays to the citrus grower as an advance. Usually, this whole agreement is not documented; therefore, there are chances of fraudulent through this whole transaction. One of the possibilities is that citrus grower might refuse to sell at the harvest stage due to high price offered by other contractors or buyers.

- **Large-sized contractors**

There are a limited number of large contractors who often purchase orchards from medium and large citrus growers. They usually possess robust finances with a purchasing power exceeding US$0.5 million. At times, they work for commission agents, wholesalers, or exporters, purchasing orchards or fruit on their behalf.

They often sell fruit to exporters and various local or inter-provincial markets based on the prevailing pricing. If prices are low, fruit can be stored in privately owned cold storages until the market price increases.

3.2.3. Descriptions of the Other Players Involved in Pakistan Citrus Value Chains

- **Commission agents**

The commission agent ('Arhti' in local language) purchases citrus fruit from producers and/or pre-harvest contractors and sells it to wholesalers /retailers/ exporters. Occasionally, the commission agents work as a wholesaler and sell directly to retailers or exporters. In some cases, citrus growers and contractors may use commission agents as a selling agent to sell in the local and far distant markets. Commission agents usually do not take the title or possession of the commodity (citrus fruit) and act as a link between buyer and seller and facilitate the whole transaction and receive a fixed amount as a commission for their services.

Nearly all the commission agents provide credit to citrus growers and contractors with the condition that they would sell their fruit to them. Usually, contractors do not receive any payments until the end of contracting period. At the end of contracting season, contractor is paid based on the agreement between the parties.

- **Wholesalers**

Wholesalers buy fruit in large quantities from commission brokers, pre-harvest contractors, or citrus growers. Wholesalers distinguish themselves from commission brokers by procuring commodities and offering supplementary services including grading, sorting, washing, and cleaning prior to selling to local markets, inter-provincial wholesalers, retailers, and consumers. Wholesalers usually extend credit to pre-harvest contractors that purchase fruit from citrus growers on their behalf. Hence, the pre-harvest contractors work as a commission agent for that particular distributor.

- **Retailers**

In Pakistan, citrus fruit is consumed fresh as a table fruit. It is mostly delivered by fruit merchants, stallholders, and street hawkers using animal-drawn carts. Fruit shops are predominantly situated in consumer marketplaces, residential areas, and along roadsides. Purchasing fruit from stores is convenient and cost-effective. Street vendors on bicycles or animal-drawn carts offer a large quantity of citrus fruit in urban and rural locations. Although they sell a little amount of fruit, they are essential in the citrus value chain by delivering the product to end consumers. Retailers, sometimes known as 'Phariwala', buy fruit directly from tiny pre-harvest contractors.

4. OPPORTUNITIES AND CHALLENGES FOR PAKISTAN CITRUS INDUSTRY

Citrus growers are shifting towards a market-oriented approach. By enhancing market-led opportunities such as crop management, improving fruit quality, and adding value through processing, the Pakistan citrus industry will grow and prosper. Newly expanded infrastructure and transportation systems will enhance the domestic and international supply chain of the citrus industry, leading to increased efficiency and reduced waste. Only 8–12% of the total citrus fruit production in the country is exported, indicating significant untapped potential for expanding citrus fruit exports. To enhance exports to European countries, the Middle East, South East Asia, China, and Central Asia markets, focus on enhancing quality management techniques, streamlining supply chains, establishing certification schemes, and reducing post-harvest losses. Pakistan's citrus business faces obstacles such as agro-ecological climate conditions characterized by severe summer temperatures, winter frost, and water constraint. Challenges in citrus value chains include inefficient production, irrigation practices, post-harvest losses, low-grade fruit, inadequate disease control, lack of fertilizers/manures in the soil, and inefficient supply systems. The citrus grower association, a private business owned by growers, is collaborating with government departments and institutions to offer necessary inputs and expertise to producers in order to enhance the quantity and quality of fruit. The shortage of proficient and trained workers for fruit picking is a marketing limitation that subsequently impacts the quality of harvested fruit. The absence of seasonal or temporary labor for fruit picking is also a challenge. Extension workers, medium, and large citrus producers are collaborating to offer essential training and accommodation to the labor force. This will reduce fruit damage during harvesting and guarantee the availability of labor when needed. Markets like Europe, the USA, and Oceania mandate adherence to Good Agricultural Practices (GAPs) and Sanitary Phytosanitary (SPS) regulations. Efforts are underway, both publicly and privately, to improve the quality of citrus fruit to meet the necessary certification standards for export.

5. CONCISE SUMMARY

Pakistan's agro-food value chain system is quite diverse, with citrus value chains mostly controlled by citrus growers, pre-harvest contractors, and citrus fruit exporters. Other key members include commission agents, wholesalers, and retailers. Citrus value chains can be categorized into two main types: unprocessed chains for local markets and processed chains for export and juice extraction. Most citrus orchards in Pakistan are smaller than one hectare, however the average area of a citrus orchard is around 20 hectares. Historically, citrus growers primarily sold their fruit or orchard to pre-harvest contractors, with only a minority engaging in direct sale of their goods in the markets. Citrus growers are now more market-oriented and are selling their products directly in national and international markets instead of following the traditional practice of selling before harvesting, thanks to the availability of market information. Few difficulties and possibilities can be tackled wisely to enhance the prosperity of Pakistan's citrus sector. The greatest potential for generating export revenue resides in exploring opportunities in the export market.

6. AT A GLANCE

Pakistan cultivates around 30 varieties of fruits, with citrus fruit being the most prominent, accounting for over 30% of the country's total fruit output. Over 90% of citrus fruits are cultivated in Punjab province and distributed through various value chains in both domestic and international markets. Most citrus fruit produced in Pakistan is consumed locally with minimal processing, although 10–12% of the total production is exported after undergoing value addition. The value chains are quite varied, involving numerous participants who play active roles in determining the final destination of citrus fruit in various supply networks. The major objective of this research is to identify several value chains of Kinnow citrus fruit in Pakistan and analyze the roles and functions of different participants in the citrus business in the country. A study of various stakeholders in Pakistan's citrus industry was done in 2013-2014 to gain a deeper insight into the citrus value chain. 245 individuals were questioned using convenience sampling during a 4–5 month period from three prominent citrus-producing districts. Citrus value chains can be categorized into two main types: raw and processed. In the past, many citrus growers engaged in preharvest contracts for their orchards, while just a few sold their orchards directly in local and overseas markets. Growers are gradually shifting towards a more market-oriented approach, leading to a change in the proportion.

Appendix A: Export of citrus (KINO) fruit, country wise

Commodity/Country	2013–2014	
	Quantity '000' tonnes	Value Thousand rupees
Kino, fresh	**353.72**	**15,665,315**
Afghanistan	112.87	4,514,597
Angola	0.09	3593
Armenia	0.00	--
Azerbaijan	0.00	121
Bahrain	1.64	57,762
Bangladesh	1.87	70,369

Commodity/Country	2013–2014	
	Quantity '000' tonnes	Value Thousand rupees
Cambodia or Kampuchea	0.27	10,368
Canada	0.73	39,012
Georgia	1.80	121,427
Hong Kong (S.A. Re. Chi)	0.19	6587
Indonesia	17.34	864,293
Iran (Islamic Republic.)	0.00	–
Iraq	0.00	–
Italy	0.00	–
Kazakhstan	0.35	14153
Kuwait	12.16	464,032
Lithuania	0.42	29,000
Malaysia	1.01	52,365
Maldives	0.00	–
Mauritius	1.45	75,519
Netherlands	0.05	2174
Norway	0.08	3419
Oman	5.79	179,048
Philippines	10.15	431,834
Poland	0.00	––
Qatar	1.77	79,950
Reunion	0.18	7321
Romania	0.07	4538
Russian Federation	72.00	436,2011
Saudi Arabia	12.63	421,972
Seychelles	0.08	5388
Singapore	2.11	87,588
Sri Lanka	4.65	157,871
Sweden	0.05	2571
Tajikistan	0.04	1172
Ukraine	21.70	1,256,946
United Arab Emirates	67.78	2,234,566
United Kingdom	1.99	86,283
Uzbekistan	0.07	2931
Viet Nam	0.34	14,533
Source: [17]		

REFERENCES

1. Ministry of Finance. Pakistan Economic Survey 2013-14. Islamabad. 2014
2. Ministry of Finance. Pakistan Economic Survey 2015-16. Islamabad. 2016
3. Ministry of National Food Security and Research. Agricultural-Statistics-of-Pakistan-2014-15. Islamabad. 2015
4. Spiegel-Roy P, Goldschmidt EE. Biology of Citrus. Cambridge University Press; Cambridge, UK. 1996
5. FAOSTAT. Download [Internet]. 2014. Available from: http://faostat3.fao.org/download/Q/QC/E [Accessed: 20 December 2014]
6. Agriculture Marketing Information Service. Crop Production Data 1947-1948 to 2009 - 2010 [Internet]. 2014. Available from: http://amis.pk/Agristatistics/production.aspx [Accessed: 20 December 2014]
7. Agriculture Marketing Information Service. District Wise Data of Citrus [Internet]. 2014. Available from: http://amis.pk/Agristatistics/DistrictWise/Citrus.html [Accessed: 20 December 2014]
8. Pakistan Horticulture Development & Export Company (PHDECo.). (2005). Citrus Marketing Strategy. Retrieved from http://www.phdec.org.pk/MktStrategies/Citrus.pdf
9. Pakistan Bureau of Statistics. Agricultural Statistics of Pakistan 2010-11 [Internet]. 2014. Available from: http://www.pbs.gov.pk/sites/default/files/agriculture_statistics/publications/Agricultural_Statistics_of_Pakistan_201011/tables/table45.pdf [Accessed: 22 December 2014]
10. Government of Punjab. Punjab Agricultural Statistics 2012-13, P. Directorate of Agriculture Crop Reporting Service, Editor. Lahore, Pakistan. 2013
11. The World Bank. GNI Per Capita, Atlas Method (current US$). 2015
12. FAOSTAT. Download [Internet]. 2017. Available from: http://faostat3.fao.org/download/FB/CC/E [Accessed: 10 January 2017]
13. Government of South Australia. PIRSA Horticulture [Internet]. 2011. Available from: http://www.pir.sa.gov.au/horticulture/horticulture_crop/citrus2/fresh_citrus_report [Accessed: 26 October 2011]
14. FAOSTAT. Compare Data [Internet]. 2014. Available from: http://faostat3.fao.org/download/T/TP/E [Accessed: 22 December 2014]
15. Chaudry MS. Opportunities and Constraints in the Production, Marketing and Export of Citrus in Punjab. Faisalabad: Faculty of Agricultural Economics and Rural Sociology, University of Agriculture; 2004

16. Ali T. Marketing of Citrus Fruit in Pakistan. Karachi, Pakistan: Department of Commerce, University of Karachi; 2004

17. Ministry of National Food Security and Research. Fruit, Vegetables and Condiments Statistics of Pakistan 2013-14, O.E.W. (MNFSR), Editor. 2015: Islamabad

18. Agriculture Marketing Information Service. Agro Based Industries in Punjab [Internet]. 2014. Available from: http://amis.pk/AgroIndustries.aspx [Accessed: 17 January 2014]

19. Ministry of Food and Agriculture. Agriculture Statistics of Pakistan 2008-09. 201

04 THE VALUE OF SHIFTING SEAFOOD SHIPPING FROM ROAD TO SEA TRANSPORT

1. INTRODUCTION: AN OVERVIEW

The Norwegian aquaculture industry, founded 45 years ago, has steadily expanded and is now a key sector in Norway's economy. Currently, the majority of fresh salmon products are delivered using roadways. This transportation method has obstacles including the risk of frequent accidents on icy Norwegian roads, poor quality of services bought in a market controlled by seafood product clients, and limited environmental sustainability. Short sea shipping (SSS) is an alternative transportation method known for producing less pollution than road transport while still offering adequate speed for market transport. This case study is on the seafood industry cluster comprising the island municipalities of Frøya and Hitra, along with the Rørvik municipality located on the coast of central Norway. The export volumes from these two industrial clusters are projected to expand fivefold. If these items continue to be transported by land, it may not only be regarded environmentally unsustainable but also result in greater road damage. The Coastal Harbour Alliance was founded in autumn 2014 through collaboration between Kristiansund and Nordmøre Harbour business and North Trøndelag Harbour Rørvik IKS. This alliance aims to increase the number of fresh fish exports by shifting freight transportation from the Hitra/Frøya and Rørvik municipalities from road to sea, utilizing two nearby ports. An essential technical aspect of this transportation system is the effective use of intermodal transport, when the same container is transported using at least two separate modes of transportation. The project primarily aims to identify a solution that meets the transportation needs of end customers in the daily seafood supply chain in terms of frequency, cost, and reliability.

This chapter further explores the value aspect of the planned freight reallocation from road to sea transport, based on a case study by Engelseth et al. [1]. This development organization is distinct from the supply chain it seeks to transform. The road transport solution considers customer preferences and the broader societal and environmental values. Value is subjective. Various parties in a supply network have differing perspectives of the value of shifting seafood freight from land to marine transport. At first, these factors are identified and talked about. Various factors affecting these value judgments include CO2 emissions, level of uncertainty encountered, accident rates, time to market requirements, product quality goals, logistical feasibility and expenses, and strength of business relationships. Viewing value as a temporary occurrence in the project development of food chains is where our investigation begins. The chapter examines this by initially exploring freight reallocation inside a networked project organization. This section describes the current state of the project within its networked environment, highlighting key network-related obstacles in reallocating freight to the SSS mode. Next, the concept of value is developed within this organizational context. Next, an introduction of SSS as a transportation mode is presented. The concluding section discusses the specific freight modes used in the case study from a supply chain viewpoint, addressing logistical and inter-organizational integration issues.

2. FREIGHT REALLOCATION: PROJECT, NETWORK AND ORGANIZATIONS

The intended shift of freight from road to sea transport is structured as a project. The potential for shifting fish transportation from road to sea is significant in this area because of the strategic position of key ports near the municipalities and the support from the government, aquaculture industry, and logistic service providers. The prevailing road transport mode now favors low-cost international trucking service providers, but the creation of a Short Sea Shipping (SSS) system with short-distance trucking feeder routes benefits local Norwegian truck companies due to their established presence in the region.

The reorganization of the SSS will require a significant amount of time. The construction of a new modern harbor facility at the industrial park on the small island of Jostenøya in the island municipality of Hitra significantly improves transportation efficiency. Seafood producers at the industrial park can directly load their products onto ships at the adjacent harbor, reducing the need for road transport from producers located a few kilometers away from the new harbor facility.

During the initial phase of the intended start-up, sea transportation will serve as a complement to road transit. The project members are currently actively searching for ways to facilitate return cargo flows. Various opportunities exist, such as the transportation of fruits and vegetables from EU countries to Mid-Norway and Western Norway that can be consolidated. The decision to switch to SSS for the new production plant, despite its proximity to the harbor, is still under active discussion within the industrial network. One primary factor is that the cost of shipping seafood exports in the region is often decided by the client. The ex works clause widely employed suggests that distant clients, who are not fully aware of production and logistical infrastructure in Norway, ultimately decide how the freight will be carried. This implies that transitioning to SSS transport for seafood requires persuading the international clients of the seafood farmers. The majority of these importers are situated in the EU and currently have their merchandise carried directly from the supplier to the recipient by trucks. Aside from the many accidents on winter Norwegian roads, this is a convenient transportation option.

Furthermore, the seafood producers and exporters are fundamentally indifferent to how their goods get transported, by which trucking company, it is only when they take into consideration societal values promoted to their own export market customers such as retailers that they may tend to prefer sea transport solutions. This, however, may also demand traceability to verify the use of transport modes. Such choices may accordingly benefit the reputation of the firms from corporate social responsibility standpoint.

Figure 1. The projected industrial park and harbour facility at Hitra.

Figure 2. Seafood on Norwegian roads is prone to frequent accidents.

The transfer of this cargo from land to sea necessitates a marketing initiative by Kristiansund and Nordmøre Harbour corporation, the owner of the industrial park and harbor facility under construction, and the Hitra municipality, which supports the establishment of this production and port service infrastructure. This challenge requires restructuring the Coastal Alliance effort to transition seafood transportation from Hitra and Frøya municipalities from land to sea transport. The main stakeholders possess a concept of value for the use of the new harbor infrastructure at Hitra. It is challenging to persuade international fish importers from the Hitra and Frøya facilities to utilize SSS for importing their products. The customer's perspective on this freight reallocation is essential. Given the several networked supply chain partners considering transit from Hitra to markets, it is crucial to analyze their varying perspectives of value. We will now analyze the definition of "customer value" in both corporate and societal contexts. We will now discuss the environmental concerns related to shifting freight to Short Sea Shipping (SSS) in this project.

3. ON "CUSTOMER VALUE" AND VALUE IN GENERAL

Value is crucial in all types of business. Value can be defined as something that people strongly appreciate, cherish, or safeguard. This suggests a semiotic comprehension of value linked to texts and discourse [2]. Within the corporate community, the emphasis is on the economic dimension of value. From a supply chain management standpoint, what is valued, treasured, or safeguarded is linked to the perceived benefits of the supply compared to the costs of ownership. The acquisition involves weighing the apparent benefits and drawbacks in a delicate balance. Christopher [3] terms this balancing in the organizational context of a supply chain as "customer value". Being a statement of subjectivity, the concept of "customer value" is inherently complex. The human mind and its preferences are no objective phenomenon; preferences are contingent and change. Value is considered in this study as always intersubjective. This is vital, because shared understanding of what constitutes customer value in the conglomerate organization of a supply chain also represents a binding force. Value is an integrator, a boundary object that helps to integrate in the supply chain. It plays the role of potential, to secure recurrent purchases through continuous customer satisfaction achievement.

Customer value is in business practice clearly a complex phenomenon. Supply chains are here considered as linkages characterized by systemic integration, an organizational structure. Being complex means that value from the perspective of the customer is seen as a process; it is emergent. Since value is always in flux, for the supplier, it is also a moving business target.

Clearly, what is written here represents a particular process approach to appreciating "value" in supply chains and thereby managing them. Given that customer value is the key objective in a supply chain and that we have already indicated that customer value is customer perception of benefits balanced against costs of acquisition and use, this in itself is regarded a process. Because, the customer evaluating follows a timeline, it is a learning process that emerges through interaction in the business relationship between the customer and the supplier. Furthermore, in line with Richardson's [4] view that no firm is an island and in line with Håkansson and Persson [5] who describe firms' interaction in supply chains as integrated hubs of interaction with various heterogeneous suppliers and customers, this entails some level of network complexity. This complexity resides at different levels of QUERY. Perrow [6] classifies levels of analysis, starting from the individual level, through group, department, division, organization, inter-organization and finally at organizational set. The "inter-organizational" level is commonly termed as the relationship. The organizational set is commonly termed as "network" or "chain" in supply chain literature, a systemic configuration of different firms working together. In this study, this network is a time-limited loosely coupled project organization bound by a common functionality of developing sustainable seafood transport. All these layers of reality in business can be considered as subject to complexity, and they do have impact on perceptions of what in practice is customer value in different ways. What is vital is having a fundamental understanding about the "value", which is dynamic in real life and can be found at different organizational levels. Furthermore, the characteristics of value at these levels interact with each other. Clearly this paves the way for a research proposition regarding what characterizes such interdependencies, but this is out of bounds of this book chapter.

In this chapter, we seek to address practical issues regarding perceptions of value associated with reallocating freight from road to SSS transport. This implies the need to simplify our analytical framework. In relation to the levels of analysis, we concentrate on customer value at the organizational and network levels as well as the relational level that binds firms together. These levels of analysis are illustrated in Figure 3.

The choice of omitting the more personal levels of analysis is because of the data, which mainly encompasses value perceptions regarding the different organizations involved in the network, regarding the research issue at hand. These perceptions are considered embedded in a network that is characterized, in line with Gadde et al. [7], by its atmosphere. A key feature of network atmosphere is the level of trust influencing the willingness to integrate.

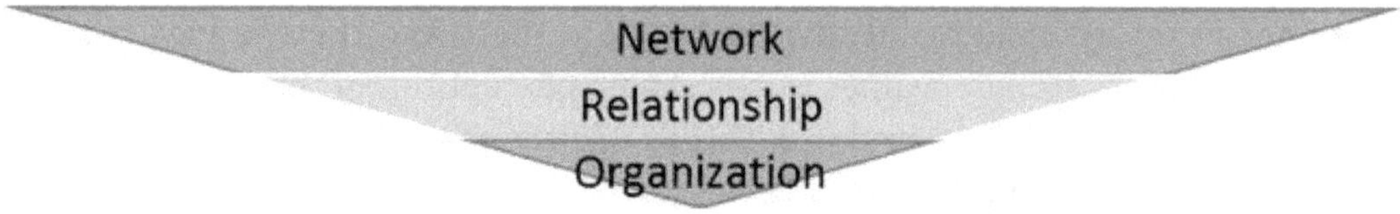

Figure 3. Levels of analysis applied in this study.

In this case, we focus on integrating a project organization associated with food freight reallocation. Describing a network as a set of interacting business relationships also entails that various relationships are interdependent. An action in one business relationship has impact on a larger network organizational structure consisting of more or less integrated firms. In this chapter, we propose a form of domino effect on other relationships, and this includes how value is perceived as emergent due to interaction in different relationships affecting the network atmosphere. Finally, it is vital to stress that, even though customer value is ultimately measured by the customer, other network actors may also perceive this factor. This is captured in the industrial marketing literature as the concept of "value proposition" [8]; value may be proposed by the supplier, and this creates foundation for dialogue to create customer value through interaction.

The value proposition implies therefore a boundary spanning effort by the supplier, aiming to reach out to interact with the customer and thereby learn first-hand what the customer needs are and is willing to pay, for a market offering. This market offering is accordingly viewed as reciprocally interdependent in the business relationship between the customer and supplier. This is clearly a process, a mutual adjustment. The customer or supplier may have carrying degrees of power, but seldom does one dictate over the other a conception of "value". This process of adjustment, learning what is valued for one's own firm and the other firm, is again embedded in a network structure.

This structure can also be conceived as having properties concerned with the meaning of "value", as discourse or alternatively termed as "network culture". Networks may, following Cooren [2], be associated with a common binding discourse. At this aggregate level, customer value is never a precisely measurable artefact. People are self-reflective, and what we like and what we prefer are in continuous change, both cognitively fluid and contingently dependent. In our case, this discourse can be described as a business culture, rules, canons and norms that define the network as a collective of firms. The network may be considered as a societal level of investigation to the degree that organizations partaking in the network are not merely business organizations. The network may also be considered as an ecosystem This implies taking into account in addition to economic concerns, also nature and societal concerns. In this study, such environmental concerns are considered contextual, not part of the network itself. The studied network does, however, consist of a mix of business organizations and public entities. Value is a complex phenomenon in a business network setting. We now turn to considering the shipping solution value is associated with.

4. SHORT SEA SHIPPING

The main common transportation modes are classified as either road, sea, air, pipelines or rail. In a supply chain, which is a business context, value perceptions will be the strongest influencer when choosing among these alternatives. From a business perspective of the individual supply chain actor, priority is given to achieve company profitability. Choices are, however, limited by the reason of distance and characteristics of goods. Recent technological innovations facilitate easy shifts between different modes possible. The container is a key resource in such transport configurations. This functionality is called intermodal transport. Furthermore, especially when transporting fresh perishable foods, reefer containers secure food quality and low-cost transport intermodal changes. Traceability is assured through temperature control that also enables maintaining a standard quality of fish products over a prolonged timeline of transport duration. The quality of traceability is dependent on the level of supply chain integration.

Intermodal transport represents combining minimum two transport modes in a particular transportation chain without any change of container; a combination of road, rail and water transport. Initial and final road transportation must be as short as possible [9]. This means that intramodality, following Boske [10], can be described as a process of transporting freight through a systemically organized network involving combinations of different modes of transport in which all the component parts are seamlessly linked and efficiently coordinated using standardized transport resources, the container being the core facility. Intermodal transport concerns investment infrastructure cost, maintenance cost for terminals, purchasing cost for vehicles and equipment and cost for transfer and storage. These factors constitute the transport process associated with intermodal transport.

The overall economic benefits of intermodal transportation are proposed by Yevdokimov [11] as divided into four elements: (1) an increase in the volume of transportation in an existing transportation network, (2) a reduction in logistic costs of current operations, (3) the economies of scale associated with transportation network expansion and (4) better accessibility to input and output markets. The cost structure for the transportation at each phase, however, is unclear and thereby hard to break down to create an accurate perception of the total that intermodal transport at an aggregate network level is cost-effective over long distances and in large volume. This is due to the inter-organizational character of this form of transport; companies reluctantly only share their cost information.

In the studied case, the supply chain undergoing potential freight reallocation, the combination of sea and road transport is advocated by the network actors with the intention to achieve efficient transportation performance. One of the reasons, as described in the introduction, is to utilize a multimodal concept with SSS as the main transport mode, to reduce road accidents, and to reduce traffic congestion in urban areas. Road driving in residence site has arose noise pollution and unsafe conditions for people in the local area. In addition to that, less greenhouse gas emission is desired in the long term from the perspective of green logistics. According to Rodrigue et al. [12], the term "green logistics" refers to logistics designed not to only be environmentally friendly but also economically functional. This implies a statement of network value, a part of the discourse this SSS development project is embedded in. There is no evidence that taking environmental considerations into logistics system would have negative influence on logistics performance [13].

Taking a supply chain perspective, managing this chain involves integration of this multimodal transportation system through a chain of interdependent suppliers. The customer, in this picture, is each intermediary, with the end user far off in the distant markets waiting to purchase and later eat their quality seafood from, in this case, Norway. Customer value, perceived by an end user in the food chain, is accordingly fragmented. It is perceived as a sequentially interdependent, a cumulative quality where each leg of transport impacts on the following. In such long-linked sequentially interdependent inter-organizational entities stretching from raw-material source to consumption, typical for manufacturing and for modernistic food production, sequential interdependencies are prominent. A following activity is dependent on the completion of a previous one; when such activities are networked between different firms, this creates interdependencies, a form of network power.

Thompson [14] describes variation in power-based interdependencies as being either reciprocal, pooled or sequential. In cases of dominant sequential interdependencies, resource pooling and intense interaction (reciprocity) support a long-linked form of production. In services, it is either core pooled or reciprocal interdependencies that are supported by sequential timing. This means that logistical coordination is a prime effort in food chains. Modernistic (industrialized) food chains are a particular form of supply, similar to manufacturing. Activities are such supply chains need, fundamentally, to be coordinated as series of events in a process. To synchronize the activities carried out in supply chain processes, a series of transport services need to be sufficiently integrated and coordinated. The aim of integration concentrates on lead time minimization and improvement on resource utilization [15]. Following Macharis and Bontekoning [9], intermodal transportation assumes optimization of its individual modes not only separate organizational transport components but also as a part of transport network as a whole system. Since this concerns inter-organizational integration, this also is a supply chain management issue.

The case of infrastructure development at Hitra to accommodate sea freight involves a particular form of sea transport sending goods from Hitra for a shorter distance, for reloading to mainly road transport, but potentially also long-distance sea transport in cases of frozen seafood transport. SSS usually defined as the shipping of cargo flows for quite short distances along a coastline. The EU Commission considered SSS as "the only freight mode that can offer a realistic prospect of substantial modal shift from road, as well as improve competitiveness and reduce environmental damage" [16]. In the current supply chain, considered as a whole system, SSS is a supplement to road transportation. Transport intermodality is accordingly a facilitator of efficient SSS. The viability of SSS also depends on type of transported freight. This consideration is generically considered when choosing the right transport mode. Air transport is fast but costly and is best suited for lightweight and smaller goods, road transport is convenient over short to medium distances since it provides door-to-door transport and sea transport is inexpensive but slow and can carry heavy bulky goods. In addition, in terms of environmental concerns, sea transport can carry large volumes in a single vessel and are therefore considered more environmentally friendly. Also road wear and accidents make SSS a preferable mode of transport compared to road transport of the goods.

To secure logistically efficient transport, SSS involves the use of technically adapted ship designs. The main technologies of the SSS are Float-on-Float-off (FLO-FLO), Lift-on-Lift-off (LO-LO) and Roll-on-Roll-off (RO-RO). The FLO-FLO ships are also known in which the floating cargo is floated into the ship's cargo space by submerging the ships' loading area. A typical disadvantage of this arrangement is that the ship must be lowered out at sea, which allows the floating vessels to be stowed into a stowage level vertically fixed within the ship. In addition, this submerging operation of the ship must be adjusted to fix deck or girder structures, which segregate the cargo containers at various cargo levels. Such intermodal transport is used mainly in developing countries with limited harbour facilities.

The LO-LO vessels, illustrated in Figure 4, are container vessels transporting a wide range of products that must be loaded and discharged in the port by crane and derricks. The cargo is lifted on the vessel according to a particular plan that is required by technical characteristics of the vessel, "not equipped with ballast-adjusting mechanisms. The LO-LO solution will be relevant for other types of cargo and included in use at Hitra. These are vessels commonly considered as "container ships".

The RO-RO technology is used for the fresh fish transportation in the studied case. Roll-on-Roll-off is the technology, which is applied in the design of ships and allows to carry wheeled cargo. This is the only one solution for sea transportation of heavy-wheeled freight such as trucks and other bulky construction and road machinery. According to Medda and Trujillo [17], the RO-RO vessels need considerable investment that implies the requirements for a satisfactory level of the commercial operations. The RO-RO vessels have built-in or shore-based ramps that allow the cargo to be efficiently driven on and off the ship during loading/discharging in the port. They can be described as freight ferries. Much as a car ferry, the ship load (in our case trucks) is driven on/off the deck on its own wheels. Advanced engineering technologies allow the ship owners to compete in the SSS market through the functionality optimization of RO-RO ships and flexibility in cargo access equipment. The use of stern ramps suitable for the different types of quays and port facilities, custom-made shore ramps, provides highly efficient and quick loading and unloading [18].

Figure 4. A FLO-FLO ship operation.

RO-RO technology illustrated in Figure 5 is primarily associated with higher speed of loading and unloading of goods compared to traditional container vessels [19]. The capacity of such vessels is unlimited compared to other modes of transport, illustrated by specialized car transport vessels used for long-distance transport. While car transport vessels are adapted to this particular type of freight, the SSS RO-RO ships are constructed as smaller vessels under 10,000 DWT adapted to reap benefits of quick loading and unloading, particularly important when ships call at numerous ports loading smaller consignments of freight.

Appling RO-RO technology implies a truck is loaded at Hitra or Frøya, driven to the port at Hitra and loaded on the ship. The truck is a semi-trailer, where the trailer section containing a reefer container remains on the vessel, while the truck returns for further engagements. Likewise, at the port of destination, a new truck picks up the trailer for further transport. The trailer has an upper side rail that is a part that is a beam that runs the length of the upper frame of the trailer. The rear reflector is a light-reflecting device that marks the back end of the reefer. The lower side rail is a beam that runs the length of the lower frame. Support legs hold the semi-trailer in a horizontal position. A piece of metal that protects the end of the support leg of the reefer is called the sand shoe. A front reflector marks the front end of the reefer using a light-reflecting device.

Much of the long-distance refrigerated transport by truck is done using such trucks pulling refrigerated semi-trailers (reefers) only onto the vessel, the decoupling form the trailer. The number of trailers involved into the transportation process is estimated to be 100 semi-trailers due to skip capacity. Return cargo is a crucial and still unsolved economic factor.

Figure 5. Loading RO-RO ship.

The project is based on the concept of collaboration between supplier, recipient, shipping company, road logistics provider and harbours. The trailer will function both as a cargo carrier and a distribution unit.

A vital factor in all transport logistics is the speed. This impacts on the time of delivery and accordingly on customer value. A negative correlation exists between vessel speed and vessel capacity. Therefore, according to Woo and Moon [13], maintaining certain vessel size is also a basis of speed control. From the terms of decision-making, managers count economic saving from slow speed and extra income which is raised by speed-up service. This shows how value incorporates in transport both cost and benefit perceptions, and this needs to be balanced. This balancing can be associated with mathematical optimization principles. However, when considering the environmental performance, slow speed is the preferable choice in maritime shipping if it is still possible to be able to meet given time limits. This entails that when widening the level of analysis to the network level, encompassing both the environmentally concerned society and the nature itself as substance matter, further perceptions on what constitutes value emerge. In the model of speed optimization on the fixed shipping routes, the main business objective is to reduce fuel consumption. This is possible to achieve by adjusting the sailing speed. Given capacity constraints of ships and harbours, as well as uncertain weather conditions, it is however difficult to optimize ship routing. Shipping is in practical circumstances, therefore a complex process where optimized routing represents plans functioning as an indicator.

5. THE ENVIRONMENTAL CONTEXT

In this section, SSS is discussed in the context of the natural environment. First a few words are provided on what constitutes environmental concerns pertinent in relation to transport mode selection. Focus is here on the factor of global warming, seen as a human-inflicted phenomenon. The processes of goods production, transportation, inventory storage and end customers' consumption are causing greenhouse gas (GHG) emissions. GHGs in the atmosphere are major foundation of the greenhouse effect. Mora et al. [20] pointed out that if the greenhouse gas emissions are not reduced, humanity can face a serious problem as the excess of the historical planet's temperature already in 2047 with its impact on ecosystems, biodiversity and the livelihoods of people worldwide [21].

After oil and gas extraction, manufacturing and mining activities, the transportation of goods is among the most polluting industrial activities to the environment and ecosystem. Road freight transport is a major source of carbon dioxide (CO2) emissions that comes from the burning of petroleum-based products in cars engines. The amount of other greenhouse gases emitted during fuel combustion is quite small. There are gases as methane (CH4), nitrous oxide (N2O) and hydrofluorocarbon (HFC) emitted during mobile air conditioners and refrigerated transport use. However, construction and renovation of port area emerge large amount of pollutants into air and water, which threaten the lives of local people and natural surroundings [22].

Reduction of emissions is an important objective of any national environmental policy. According to the international Kyoto agreement, Norway is aimed to reduce greenhouse gas emissions to 84% of the registered level of emission in 1990. There is also an agreement implied by Norwegian parliament in 2008, where the emission reduction is set as a national target. The Ministry of Transport is working to achieve these goals by stimulating public transport and innovative types of transport, by investment in new technologies and encouraging a change to vehicles with lower emissions. This policy is also supported by the economic instruments: CO2 taxes and green energy subsidies. These policies are intended to promote transition to increasingly using environmentally friendly transport modes.

Road transport involves consuming unclean fossil fuels, which emits CO2, NOx, sulphur compounds and other toxic substances. These substances seriously damage air quality, also at a local level. It is a serious health concern, especially for people with breathing disorders. Rommert et al. [23] analyse choice of transportation (transportation mode, intermodal transport, equipment choice and fuel choice) against their environmental impact. This study reveals that larger scale of cargo to transport will lower CO_2 emissions per g/t/km. Among all transportation modes (container vessel, rail, truck and plane), transport by vessel is the most CO_2 efficient and offering highest fuel economy. It is a price-wise and economical transport mode. Therefore, the SSS is attractive from an environmental perspective in likewise as being economically feasible [16]. This form of transport usually does not allow door-to-door delivery and is slow.

Governments are not sitting idle while ships pollute the air. Sea transportation is thus subject to a particular tax system. The Sulphur Emission Control Areas (SECA) for shipping companies was entered into force in September under new EU legislation. There have been introduced certain taxes and requirement to use fuel where the sulphur content should not exceed 0.1% when operating within the SECA. The SECA includes North Sea, which Norwegian shipping companies are crossing during their import/export operations. Therefore, it will have impact on operation cost and will increase charges to the customers [24]. Ships emit NO_2, but this transport-related value can be improved based on the available or a new technology:

- Fuel switching
- Development of new technologies and vehicles that are more efficient
- Minimization of fuel use by adopting driving practices
- Improvement on maintenance
- Switching from one transport mode to another during a transportation chain

In our case, all transportation related to export of fresh fish to the EU countries represent about 130 mill.ton.km/year with emissions equal 76 g CO_2-ekv./ton.km [25]. Sustainable transportation is a transportation that satisfies individuals and society's needs in a long-term perspective. This implies production sensitive to human and environmental health concerns, in equal conditions for current and future generations. Sustainable transport is economically efficient and energy effective, competitive, operating offering alternative transport solutions and accomplished by the use of innovative technologies. A sustainable transport system considers environmental aspects and aimed to reduce emissions and waste, minimize consumption of natural rare resources and decrease the use of land and production of noise. In addition to this, there is a strong transport link between sustainable transport and reduced accidents and congestion on roads. Sustainable transport contributes to economic development. Social progress and living quality are improved by implementing the concept of sustainable transport. Thomaeus [26] points out that a transport system can be concluded as sustainable where economic efficiency and environmental protection complement each other or in a balance with acceptable level. The adjusted system of information sharing and integration is vital for sustainability of fresh fish supply as perishable goods and greatly influences the waste reduction. The estimated food waste from manufacturer to the end customer in different supply chains is about one-third from the production volume. Some major causes of waste in fresh fish supply are weather conditions, lack of coordination of supply flows, road accidents, failures in the transportation process, shortcomings in the shelf-life management and inconsistency between demand and supply.

Taking a sustainability perspective, it is imminent to include considerations of food waste. This is because much of the seafood transported from Norway is in a fresh state making them perishable goods. Poor transport quality increases the risk of food waste during transport impacting on ethical considerations regarding efficient food use in relation to human welfare aimed at a global level. Food waste reduction involves fundamentally improved coordination of food production volume with customer demand and improvements on the efficiency and performance of the supply chain as a whole system. In the case of fresh seafood supply, it is necessary to consider the specificities of transportation and features of the product, affecting management and performance of the whole chain [27]. The logistics providers in fresh fish supply chain targeted to deliver the product to the end customer in perfect condition and maximize available shelf-life time. Seafood that is poorly delivered indicating transport discrepancies may be improved by changing the design and use of the products. Three main characteristics of the food market, affecting the structure of the supply chain, were identified by Kittipanya-ngam et al. [28]:

- Demand uncertainty. Customer demand is influenced by natural factors as weather conditions and seasonality and encouraged by promotion actions [29].
- Customer order lead time. Usually, lead time required by customers is short.
- Supply chain lead time allowance meaning perishable goods are characterized by limited life cycle and efficiency in the supply chain; allocating the sharing of the expected time to transport and handle goods between the logistical elements of supply chain is of crucial importance.

The commodities' perishability prevents the creation of an inventory buffer to account for changes in demand and transportation difficulties. According to Ahumada and Villalobos [30], this issue can be addressed through supply chain flexibility and enhanced velocity. Collaboration among chain players at an operational level, supported by integrated systems and advanced forecasting methodologies, enables the attainment of the necessary speed and flexibility. This could enhance logistical cooperation within the supply chain. The case study focuses on analyzing the fresh fish supply chain, which includes the producer, port of loading, logistics provider, port of discharge, and end client. We will now analyze the economic aspects of this conglomerate organization, the network under study, and its affiliated organizations.

6. SUPPLY CHAIN AND LOGISTICS

Customer-responsive supply chains, the "value chains", have the capacity to combine scale with product differentiation, and cooperation with competition, to achieve collaborative advantages in the marketplace. It makes commitments to the welfare of all strategic partners, including appropriate profit margins, fair wages and long-term business agreements, balancing cost with benefit perceptions. It emphasizes high levels of performance and inter-organizational trust. A supply chain thus consists of a system for sharing information among strategic partners, including shared values and vision, shared information and shared decision-making [31]. A value perspective of the supply chain as unit of analysis for transport mode alteration, accordingly, places focus on value perceptions of the supply chain collective studied in this case regarding reallocation of freight to the potential of SSS.

Economic matters in such a supply chain structure can be measured by service level, and this metric is affected by the factors like time to market, product quality, customization and flexibility [3, 7, 15]. Efficiency is then measured by calculating the economic saving of transportation cost and positive contribution on environmental protection. Despite advantages of supply chain integration, decision-makers always balance saving on supplier integration and relative cost raised by it, the transaction cost factor. Perols et al. [32] describe two types of supplier integration having paradoxical impacts on time to market. Time to market can be accelerated by supplier process integration while product integration slower time to market. And it also reveals that a positive technology spillover effect occurs with successful assimilation within a strategically organized supply chain.

The main participants of the freight reallocation project are:

- Municipalities: Hitra and Frøya
- Members of The Coastal Harbor Alliance: Kristiansund and Nordmøre Harbour IKS with Hitra Coastal Port as a part and Vikna port authority Rørvik Harbour KF
- Producers/aquaculture: Marine Harvest, Lerøy and SalMar

- Road transportation companies: DB Schenker, Bring and OTTS
- Shipping companies: Blue Water

Their role in the seafood export chain varies. Some of these networked actors are directly involved in goods handling and/or ownership; others, such as the Hitra municipality, are only indirectly linked within this chain. When the projected SSS is in operation, these supply chain actors need to cooperate together to detect and reduce environmental impact, the carbon footprint and pollutions upon the whole supply chain and gain economic benefits from their somewhat detached and local network perspective. This represents the operation perspective, a continuous network effort. It is expected that the production of aquaculture will grow to 800,000 tons by 2020. Production will be five times larger than the current production volume in 2050. According to long- and short-term forecasting, a considerable increase in the production of seafood and other goods is estimated and will require new transportation solutions that will reduce CO2 emissions caused by trucks, road accidents and road maintenance coast. New automotive technologies, improved fuel, development and improvement of the road system promote a sustainable growth of road transport share in transportation. All export flows of fresh salmon are transported by road between South Norway and the EU markets. Salmon road Fv. (regional highway shown in Figure 6) 714 is connecting the coastal municipalities Snillfjord, Hitra and Frøya, Orkanger and Trondheim. The 57.6-km-long stretch between Haugen in Orkdal municipality and Sunde in Snillfjord Municipality has low standard. The daily goods flow on trucks loaded with salmon lies between 50 and 80 semi-trailers per day. This corresponds to 17,000 semi-trailers from Hitra/Frøya yearly. Sixty percent of them are oriented to the EU markets. Traffic figures from Nord-Trøndelag are more than 3000 per year. Considering together these volumes, there is a possibility to reduce the number of semi-trailers between Mid-Norway and Europe for 12,000–13,000 vehicles not considering the return cargo flows. The use of foreign transport logistics companies has proven to increase competition and the number of accidents on Norwegian roads considerably. During winter period, transportation becomes especially challenging. Such conditions are an important reason for development of the terminals and harbours of intermodal transportation in this supply chain.

The official opening of Hitra Port took place on 16 October 2014. Regular container ship calls started in November 2014. After 5 years, the main elements of the infrastructure are on the place. The infrastructure of the port includes production and social components as engineer communication, gas, electricity and water supply system with huge water reservoirs and waste water solution. HirtshalsHarbour in Northern Denmark was initially considered a starting point for the establishment of a sea transport connection. The sea transport solution will reduce current cost of road transportation by 20–25%. The price depends on volumes of return cargo flows that will be obtained over some time. Government support and financing are of high importance in implementation phase. The calculations that have been done show that an increase in return cargo flows by 10–50% can reduce prices about 10–45% depending on the distance from receiving port to the end destinations [25]. Handling equipment, communication lines, two new aquaculture plants, warehouses, facilities of the companies providing service and maintenance for aquaculture and marine industry and another buildings and facilities are included into the project and will be built after some time.

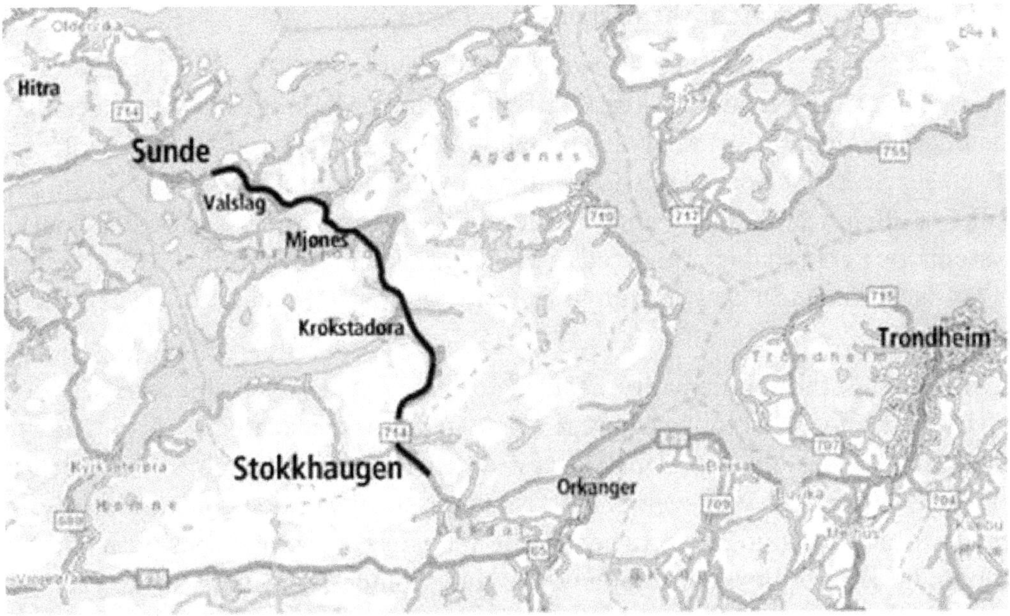

Figure 6. The "salmon road".

Hitra Port is located right in the fairway between Trondheim and Kristiansund and is therefore commonly considered a natural traffic and logistics hub for seafood and fishing industry in the Hitra/Frøya region. Hitra Coastal Port and its underlying commercial space, Hitra Industrial Park, total represent a development area of around 1.5 million m2 (1500 da). They are labelled as a "seafood logistics centre". The seafood logistics centre is directly connected to the main origins of seafood production in Mid-Norway. A well-connected transport network and extensive logistics capacity makes it possible to manage further increase in future seafood transport demand. The salmon production industry is growing steadily. Indeed, it exhibits great opportunities for cooperation with the EU markets and excellent possibilities to service Mid-Norway and Northern Norway. Many shipping companies and transportation companies are interested in using the Hitra Port as both a seafood and general cargo/unit loads hub, storage hotel, a regional distribution centre, trans-shipment terminal, hub for speed boats and ferry passengers, special storage, etc. Several companies have expressed interest in establishment in the area; some companies are in negotiating position. The world's largest salmon producer, Marine Harvest, has now secured 50 da (+ option for another 10 da) to build new salmon factory in this area. There will also be good opportunities for Hurtigruten (the daily coastal liner carrying both light freight and passengers) and cruise vessels in the port. Hitra municipality will establish a future-oriented and sustainable environment. The convenient location, along with great and new quay and harbour facilities, will provide great opportunities for economic development in the region and within the company. This indicates a multiple use potential for the harbour facility at Hitra.

Much infrastructure has already been developed at Hitra during the last 2 years. The cargo terminal Hitra Coastal Port is operational from this year and includes terminal facilities, areas and equipment appropriate for both RO-RO and LO-LO services. The port's logistics centre provides possibilities for frozen and cold storage, offers warehousing and transit storage and provides a good distribution systems via Fv714 that is connected to the port through Hitra tunnel. The upgraded Fv714 is 75 km to Orkanger and 40 to Frøya and is conducive to efficient cargo distribution.

Originally, the municipality decided to build the 120 acre industrial area of north-west side of Jøstenøya where the industrial park is located at Hitra. The first plots are already sold to Marine Harvest AS and Brødrene Sunde AS that secured themselves the land for development. Now also Bewi, Sunde Group and Lerøy are in the course of building factories at the industrial park adjacent with the two salmon factories. The Bewi company and their competitor Sunde Group each will build a new factory for the production of polystyrene boxes—fish crates factories. They do this in order to increase capacity and to be even closer to the aquaculture companies that also will establish themselves in the region. This will contribute to greater security of supply for Bewi's customers in the region. Bewi considers that it is important to be established in an industrial hub that Jøstenøya will be. The overall strategy of Bewi and Brødrene Sunde is to be a supplier of packaging for both aquaculture and agriculture and building industry. This implies they are competitors in the clustered location. Their overall logic of having production facilities at the industrial park is to provide their customers even better in quality, flexibility and environment through their innovative and trend-setting products, by being near them. The juxtaposition of the packaging factors implies considerable investment leading to logistics efficiency. Keep in mind that the volume of packaging freight, since the type of packaging used may not be folded or made smaller in any way for transport, equals the volume of packaged goods freight. By building of a new factory, Lerøy plans to merge factories in Hestvika and Dolmøya to one factory in Jøstenøya. These two factories work reasonably well today. The company Lerøy has been working for several years to merge the two factories and make the production process more efficient. This implies increase production efficiency for this company as well as logistical improvement since the two factories are now to be co-located by the harbour.

Due to demand by the companies involved in establishing themselves in the industrial park at Jostenøya, the industrial area must expand eastwards with another 60 acres. The cost of the expansion is estimated to 46 million NOK; the municipality finances it by borrowing (through a loan). It is important that companies build their facilities quickly, so that the municipality can gradually collect tax on the investment in Jøstenøya. The Hitra municipality has a strong position and extraordinary potential for industrial growth in national context. It may be possible after meetings with the Ministry of Transport where the project was presented to apply for start-up support for fresh fish exports from Mid-Norway. It is given that the full power of a restructuring of the transport side must happen this year.

These descriptions reveal how value in the supply chain clearly is differentiated between the actors, and that value is networked. In addition, through networking what constitutes a good, valued solution becomes an emergent outcome of this type of exchange process. The discussion also is interesting because it does not reveal choice of transport mode as on the agenda, even though the industrial park also is designed as a harbour. It is therefore a clear implication that this is something the Coastal Alliance needs to market to the companies establishing themselves in the industrial park. This is because it is through the business relationships between the seafood exporters and their customers that the choice of transport mode is decided upon.

7. CONCISE SUMMARY

The study's key discovery is that the idea of "supply chains" depends on different developmental programs. This study's design explicitly focuses on the problems related to reallocating freight from road to short sea shipping (SSS) without providing a detailed illustration of the seafood supply chain. This chapter's narrative discusses the features of the development project as appropriate. This project organization is somewhat flexible. The origin of the actions to secure SSS freight of fish by a political entity, which this group represents, is uncertain and it is questionable if they would ever stop. It is challenging to assess this initiative as either "inefficient" or "effective" due to its nature as a political organization. The statement reflects social value issues, encompassing environmental worries of the general public, and economic interests of the corporations participating in the project. The perceiver holds value, which is a variable factor. During the inquiry, the researchers also faced incidents that interrupted this investigation. Actors became more discreet on stage due to a logistics service provider's perception of the central Norwegian government's lack of support for the reallocation endeavor. An understanding of the supply chain as a culturally ingrained activity became apparent. These perspectives are crucial when considering the importance of "value" in both the development project and the economic significance of salmon supply, which is the project's goal. This suggests that value is not solely dependent on the perceptions of different players in the developmental network, but can also be influenced by the acts of a single actor. The project's value is interconnected and mutually dependant. Individual networked players do actions that result in changes noticed by other actors, prompting them to respond and adapt to the altered circumstances. Interacting through a network causes actors to reconsider their actions in the development process. This strongly suggests that "value" in the supply chain is a dynamic phenomena. This remark suggests that we view the developmental project as a supply chain with its own degrees of integration and issues related to trust and coordination, distinct from the supply chain that the freight reallocation initiative is intended to affect. Customer value has become widely used in supply chain management literature. It is closely associated with achieving a supply chain that is responsive to client needs. Enhancing consumer focus and awareness by utilizing adaptable resources involves establishing a responsive supply chain. This may be viewed as a source of competitive advantage. This study examines the value resources inside a specific food chain across several spatial and temporal dimensions, challenging the naive perspective of a static, space-oriented approach.

The study reveals that "value" in a supply chain setting is a dynamic and complicated phenomena. Various facets of this complexity are uncovered in this investigation. Value is interconnected. Initially, value varies concerning both the supply chain and the developmental project. This study does not pertain to the supply chain of seafood export. Second, the supply chain is a network consisting of differentiated actors, a source of both complementarity and conflict. Geographically, this space dimension creates barriers to close personal engagement and allows for varying ideas of value among interconnected supply chain actors. Supply chain management emphasizes the integration of supply chain players based on a similar goal to match value perceptions. However, we challenge the effectiveness of this approach. Increased similarity among enterprises leads to more friction and raises the risk of paradoxical occurrences. Pooling similar actors in this project organization is particularly tough. They are prospective competitors.

Diversity within a network provides valuable differences that can enhance each other and drive innovation. The dissemination of value perceptions prompts us to acknowledge that client value may not always surpass other value concerns. Customer value is often viewed as a top-level value in supply chain academia and practice, seen as a clear and unchanging purpose. This perspective overlooks the potential for supply chain actors to engage in a beneficial learning process, where customers can learn from suppliers to enhance the value of the supply chain. Connectivity also involves the privilege to retain confidential information, integrated inside an evolving project discussion. This brings us to the third and last observation. Value is a constantly changing occurrence. This modification feature relies on high-quality interconnections within the supply chain. This suggests that integration is a crucial endeavor aimed at connecting individuals to facilitate communication and learning, rather than just harmonizing express ideals. Value is subjective, and as a result, it is ambiguous in the case description. This suggests that maintaining a supply chain can be seen as a complicated process, potentially a system. A systems approach is possible when development aims to unify the network based on a shared and agreed-upon declaration of supply purpose.

The primary suggestion for the developmental project is to provide a network environment that emphasizes trust and learning. Facilitating integration involves planning and consistently enhancing the interconnections among the project participants. The primary focus of development should be on building connectedness, trust, and values considered as emerging network features, rather than immediately searching for common values. Value has been demonstrated to arise in this strategic project network. The journey is challenging and there is a possibility of failure. We suggest enhancing integration efforts to a more advanced degree of strategic thinking in project design. According to this perspective, there are no assurances on the output features. Emphasizing the procedure over the end result undoubtedly carries hazards. Additional research could focus on investigating the development of supply chains from a process-oriented perspective. This study can be structured using a case study research approach, with action research being a key component of the study.

REFERENCES

1. Engelseth P, Karlsen IV, Huang S and Hoff A. Network Constraints of Reallocating Seafood Freight from Road to Sea. Operations and Supply Chain Management. An International Journal. Forthcoming. 2017
2. Cooren F. Organizational Discourse. Cambridge UK: Polity Press; 2015. 99
3. Christopher M. Logistics and Supply Chain Management. London: FT Prentice Hall; 2016
4. Richardson GB. The organization of industry. The Economic Journal. 1972(September):883-896

5. Håkansson H, Persson G. Supply chain management: The logic of supply chains and networks. The International Journal of Logistics Management. 2004;15(1):11-26

6. Perrow C. Complex Organizations. Brattleboro VT: Echo Point Books & Media; 2014. 194

7. Gadde, LE, Persson G and Håkansson H. Supply Network Strategies. Chichester UK: Wiley; 2010

8. Kaplan DP and Norton P. Strategy Maps: Converting Intangible Assets into Tangible Outcomes. Cambridge MA: Harvard Business Press; 2004

9. Macharis C, and Bontekoning YM. Opportunities for OR in intermodal freight transport research: A review. European Journal of Operational Research. 2004;153:400-416

10. Boske LB. Multimodal/Intermodal Transportation in the United States, and Latin America-Government Plans, Policies, and Programs. Austin: LBJ School; 1998

11. Yevdokimov YV. Measuring economic benefits of intermodal transportation. Transport Law Journal. 2000;27:439-452

12. Rodrigue JP, Comtois C and Slack B. Green logistics (the paradoxes of). In: Brewer AM, Button KJ and Hensher DA, editors, The Handbook of Logistics and Supply-Chain Management. London: Pergamon/Elsevier; 2001

13. Woo JK and Moon DSH. The effects of slow steaming on the environmental performance in liner shipping. Maritime Policy & Management. 2014;41(2):176-191

14. Thompson JD. Organizations in Action. New York: McGraw Hill; 1967

15. Persson G. Achieving competitiveness through logistics. The International Journal of Logistics Management. 1991;2(1):1-11

16. EU Commission. Strategic Goals and Recommendations for the EU's Maritime Transport Policy until 2018. Brussels: Commission of the European Communities. 2009. Available from: http://eur-lex.europa.eu/legal-content/EN/TXT/?uri=URISERV%3Atr0015/ [Accessed 23-01-2015]

17. Medda F, and Trujillo L. Short-sea shipping: An analysis of its determinants. Maritime Policy & Management. 2010;37(3):285-303

18. MacGregor. Shortsea RoRo Ships. 2016. Available from: http://www.macgregor.com/en-global/macgregor/merchant-ships/shortsea-roro-ships/Pages/default.aspx/ [Accessed 14-03-2016]

19. Todorov DM. Ro-Ro Handbook: A Practical Guide to Roll-On Roll-Off Cargo Ships. Atglen, Pennsylvania, USA: Schiffer Publishing; 2016

20. Mora C, Frazier AG, Longman RJ, Dacks, RS, Walton MM, Tong EJ, Sanchez JJ, Kaiser LR, Stender YO, Anderson JM, Ambrosino CM, Fernandez-Silva I, Giuseffi LM and Giambelluca TW. The project timing of climate departure from recent variability. Nature. 2013;502:183-184

21. Statistics Norway SSB. Aquaculture, 2014, Final Figures. Continued Growth in Aquaculture, Figure 1. Available from: http://ssb.no/fiskeoppdrett/ [Accessed 03-12-2016]

22. Rondinelli D and Berry M. Multimodal transportation, logistics, and the environment: Managing interactions in a global economy. European Management Journal. 2000;18(4):398-410

23. Rommert D, Jacqueline B and Ioannis M. Operations research for green logistics: An overview of aspects, issues, contributions and challenges. European Journal of Operation Research. 2012;219(3):11100-11111

24. DSV. Global Transport and Logistics. EU Environmental Low. Sulphur directive (SECA). http://www.ie.dsv.com/about-dsv/press/News/2014/11/EU-Environmental-Low-Sulphur-Directive/ [Accessed 05-05-2015]

25. Netter JE. Sustainable Transport Solutions for Fresh Fish from Central Norway to the Continent. Final Report. Hitra Municipality. Trondheim, Norway: Enova Project; 2013

26. Thomaeus B. The European Concept of Green Corridors in Sustainable Transport Systems: Case of East-West Transport Corridor in the Baltic Sea Region. United Nations Framework Convention on Climate Change. Kyoto Protocol. 1998. Available from: http://unfccc.int/kyotoprotocol/items/2830.php/ [Accessed 25-03-2016]

27. Kaipia R, Dukovska-Popovska I and Loikkanen L. Creating sustainable fresh food supply chain through waste reduction. International Journal of Physical Distribution & Logistics Management. 2011;43(3):262-276

28. Kittipanya-ngam, P, Shi Y and Gregory MJ. Food Supply Chain (FSC) in manufacturing companies –an exploratory study on product and configuration. In: Proceedings of 17th EurOMA Conference; Porto, Portugal; 2010

29. Taylor DH and Fearne A. Demand management in fresh food value chains: A framework for analysis and improvement. Supply Chain Management: An International Journal. 2009;14(5):379-392

30. Ahumada O and Villalobos JR. Application of planning models in the agri-food supply chain: A review. European Journal of Operational Research. 2009;196(1):1-20

31. Stevenson S, and Pirog R. Values-based food supply chains: Strategies for agri-food enterprises-of-the-middle. Annual Meeting of the Rural Sociological Society; Manchester, New Hampshire. July 28. 2008. Available at: http://www.allacademic.com/meta/p244246_index [Accessed 05-05-2010]

32. Perols J, Zimmermann C and Kortmann S. On the relationship between supplier integration and time-to-market. Journal of Operations Management. 2013;31(3):153-167

05 AGRICULTURAL DIVERSIFICATION IN JAPAN

1. INTRODUCTION: AN OVERVIEW

Japanese agriculture has seen various challenges in recent decades, including low profitability, labor shortages, farmland abandonment, and decreased competitiveness against low-priced imported goods.

Japanese modern agriculture was historically defined by certain agricultural policies implemented by the General Headquarters (GHQ) that oversaw Japan immediately following the conclusion of World War II in 1945. GHQ implemented two significant measures to overhaul Japanese agriculture.

- Emancipating farmland from rich farmland-owners and re-allocating it to poor peasants in order to activate Japanese agriculture and improve food production of Japan, which was much damaged during the war.
- Forming agricultural cooperatives [1] to advance the supply chain of food industries in Japan, that is, establishing Japanese agricultural cooperatives (JA) [2, 3] in 1948 based on Agricultural Association that was established during the war to control national food supply.

The farmland emancipation program led to the emergence and expansion of numerous independent farmers, boosting their drive for agricultural development and enhancing food output in Japan during that period. JA served as a proficient food supply network for the Japanese population. However, after a long time passing, these two policies resulted in two serious issues for Japanese agriculture.

- The farmland emancipation policy resulted that the average level of the Japanese farmers' production was lowered because the average level of the farmland area was lowered. The small farmland area also obstructed modernization of agriculture by inducing cultivating machines. The difficulty for inducing modern machinery system into agriculture caused low productivity and high production cost of agriculture in Japan comparing to that in other countries. According to the data of year 2010 by the Ministry of Agriculture, Forestry, and Fisheries (MAFF) in Japan, the average farmland area per one farm in Japan is around 2 ha, that is almost 1/100 of that in the USA. Therefore, in the progress of globalization and trade liberalization on food market, Japanese agriculture has been losing their competitive competence.

- JA has been grown to a huge organization along the growth of Japanese economy, as indicated by the fact that the number of the total members of JA grew over 10 million in 2014. The business area of JA was enlarged to cover wide area regarding farmers' daily lives, such as banking, insurance, sales of seeds, sales and lease of cultivating machines, managing super markets, managing gas stations, publishing newspapers, postal service, and so forth. Particularly, their power on farmers' banking became very large as indicated that they are now dealing the amount of savings around a hundred trillion yen (circa a trillion US dollar). By the growth of JA, their organization has changed to a large bureaucratic one and lost the flexibility as a function of supply chain in agriculture. Because the distance between farmers and consumption market became large in the huge bureaucratic supply chain system, sometimes it became difficult for farmers' products to fit the market demand.

The Japanese food industry demand has evolved in response to changes in Japanese society post-war. Japanese people's food preferences shifted from traditional Japanese cuisine to Western/American foods, from homemade meals to processed food and dining out, and from family meals to eating alone. Changes in the Japanese food market have led to a shift in the types of agricultural products in demand. However, many Japanese farmers have struggled to adapt to these changes due to a lack of understanding of the market dynamics within the traditional agricultural value chain. The intricate supply chain network in Japanese agriculture involving several intermediaries such as JA, retail outlets, family restaurants, and food processors, has reduced the profits of farmers. Their sales in the Japanese food business account for less than 20% of the overall food market transactions. Figure 1 roughly illustrates the flow of food material in Japan. The low productivity and profitability of Japanese agriculture have led to a trend where the younger generation is leaving farms to work in manufacturing factories or other businesses in urban areas. This has caused depopulation and an increase in the average age of rural districts. As a consequence, regional agriculture suffered from unfavorable repercussions, leading to a shortage of manpower and the abandonment of fields in such areas. Recently, the Ministry of Agriculture, Forestry and Fisheries (MAFF) in Japan implemented a program of agricultural diversification to address challenges in Japanese agriculture. The concept is occasionally referred to as "sixth sector industrialization." The term's connotation involves integrating primary industry (agricultural production), secondary industry (processed food manufacturing), and tertiary industry (food retail and restaurants) through the reallocation of a farm's commercial resources and assets. Put simply, farmers are encouraged to adapt their products to market demands and increase their profits by expanding their involvement in various aspects of the food industry value chain. This involves reallocating resources and integrating new components into the agricultural value chain. However, the strategy faced challenges due to the lack of familiarity among Japanese traditional farmers with retail service and processing industry. Most of them lack the ability to diversify or integrate their value chain.

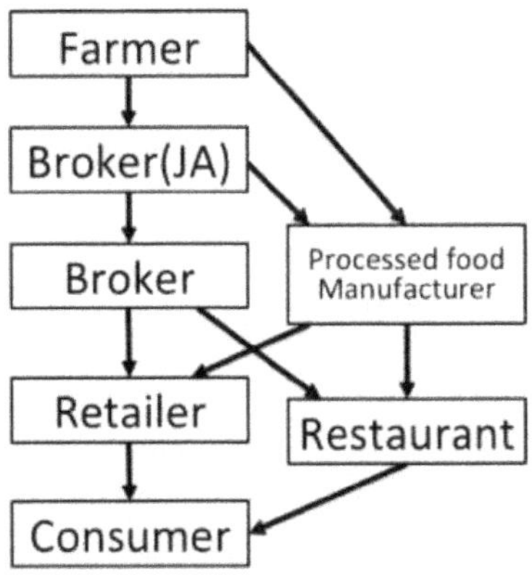

Figure 1. Value chain of agriculture food industries in Japan.

This chapter outlines various successful agricultural diversification trials in Japan to derive insights on the Key Factors of Success (KFS) in agricultural diversification. (The author has already presented certain situations in various contexts [4, 5].) If certain implications prove useful in other regions or countries, this study could provide valuable insights to enhance global agriculture.

2. FRAMEWORK OF RESEARCH

This chapter describes totally seven cases of actual diversification in Japanese agriculture to obtain implications on KFS.

The cases are classified to four categories by the difference of the key persons for establishing a new business, as follows.

1. Cases of diversification by brokers.
2. Cases of diversification by processed food manufacturers.
3. Cases of diversification by retailers.
4. Case of diversification by business persons or enterprises.

The research focuses on the following standpoints.

- Motivation and capability of key person for business
- Relationship and collaboration between farmers and key person
- Some other conditions for diversification, such as business strategy on branding/promotion, market categories, market size, market characteristics, and so forth

Based on these case studies, KFS of diversification will be discussed. For each case, direct interviewing and hearing to the key persons were effective to clarify detailed background and conditions for business success. Particularly, it was too difficult to get information on key person's personal motivation for business and relationship between key person and local community, only by literatures and public materials. Direct talking with those persons was very effective in this aspect.

3. CASES OF DIVERSIFICATION BY BROKERS

3.1. Case of Mr. RishoAzechi and the Villages Along Shimanto River

The first case of key person for agricultural diversification is Mr. RishoAzechi. He was working for a local JA office at first when he was young, and then, he became a staff of an enterprise (a third sector, a semi-public corporation) for encouraging local villages along Shimanto river. The enterprise was named as "Shimanto Drama." The whole staff members of the enterprise were only two persons at first. After that, he became a CEO of the enterprise. His first mission was to encourage the local agriculture of these small villages in mountainous area. But the agriculture products of these villages were not characterized with special advantages. Their products were ordinary regional items such as local chestnuts, mushrooms, river shrimps, and so forth.

One important event for advancing his business was the encounter with a certain designer, named Mr. Makoto Umebara. Mr. Umebara was a very unique designer focusing on package design and catch copy for local products. His first remarkable work on local products was the designing package and making a catchy on bonito, fish. There was a fishery enterprise faced on the crisis of bankruptcy. They were involved in traditional fishing using rod. Their fishery product level was too small compared to that of large enterprise with fishing using net, even though their fishes' quality was good without damages by net. They requested Mr. Umebara to design the package of a box for carrying bonito fish. Therefore, Mr. Umebara came up with a catch copy for the product as "Professional fisherman fished it, and he cocked it in the best way." His design of the package emphasized the primitive taste of fishery by rod with simple color construction. By his copy and package, the item suddenly became popular and could be sold with higher price than that of ordinary bonito by net fishing. That is, he successfully established a brand of rod fishing.

Mr. Azechi heard about the fame of Mr. Umebara and asked him to advertise the local products of Shimanto Drama. However, Mr. Umebara's strategy was not direct advertising the products. First, he published a book cheering the wonder of the lives in riverside. The title of the book was just "Water," and the content of the book consists of many essays written by many famous, eminent persons on politics, art, literature, illustration, and so forth. Generally, it is not so easy to request those persons to write an essay for a book for such small local villages and a tiny enterprise. But Mr. Umebara asked it to them without any special connection just only by saying "If you could accept our proposal, we would like to gift you a lot of river sweet fishes for a year as a sign of appreciation, instead of money." This unique approach led a success, and around 30 prominent persons wrote an essay for them.

Mr. Umebara also recommended to issue a magazine tilted "Lives in river-side." They established a membership club to love lives in riverside and issued a magazine provided to the members. The contents of the magazines were the description of the wonder of lives in river-side in local mountainous area.

These activities were thought to be a kind of branding the villages along Shimanto river. Gradually, their products were getting popularity with the Shimanto brand that has an impression of "slow lives," "sustainable life style with nature," or so forth.

The relationship between regional farmers and Mr. Azechi is very tight and close. Mr. Azechi contributed to the region through his business of retailing the Shimanto products. In that sense, even though the regional farmers were not familiar with marketing and branding, Mr. Azechi in behalf of them established business for them.

Now, Mr. Azechi is also managing Michi-no-eki, a local shopping store including restaurant for regional agriculture products, established by the government for encouraging regional farmers. At first, the local government opposed to build it in the rural area where Mr. Azechi proposed because of low traffic level. However, after once they built a shop there, over a hundred thousand people visited there for half of a year in terms of the charm of the shop.

3.2. Mr. Mochifumi Toutani and Citron Business in Umaji Village

Mr. MochifumiToutani was a leader of local JA organization of a rural village, named Umazi (meaning "a road for houses"). The village was located in mountainous area, and their population was only about one thousand. They were engaged in forestry industry in the past. However, by globalization, their products lost the competitive competence comparing to low price materials imported from emerging countries. They lost the income through globalization. As they could not depend on forestry, they initiated a new business by utilizing citron (Yuzu, a kind of oranges) that they were incidentally growing in the forest. However, most of their citrons were deformed and not suitable for selling as a fruit item. So that, they made juice by processing citrons. But it was also very difficult for them to sell their juice items because there were lots of similar products in the market. Mr. Toutani then focused on marketing and promotion on their citron juice. They emphasized that Umaji village was a rural, remote village but was plentiful of nature, and the people's personality in Umaji is primitive and friendly. After the promotion effort of over 20 years, their village name became famous and the brand of Umaji, accompanied the image of natural lives in mountainous area, was proliferated.

Particularly, improvement of Internet and nation-wide delivery service gave them advantage in business. Nowadays, they were selling their citron products including juice, processed seasonings, spices, and so forth through Internet and delivery service. They have their own factory for processing citron, and the factory is opened for tourists as a kind of tourism service. They have a restaurant in the factory and hold some kinds of attractions for gaining tourists for the village. Their annual sales reach up to 3 billion yen (around 30million US dollar) with the profit rate of over 10%. Growing citron and processed food industry on it are now their main businesses to support the village.

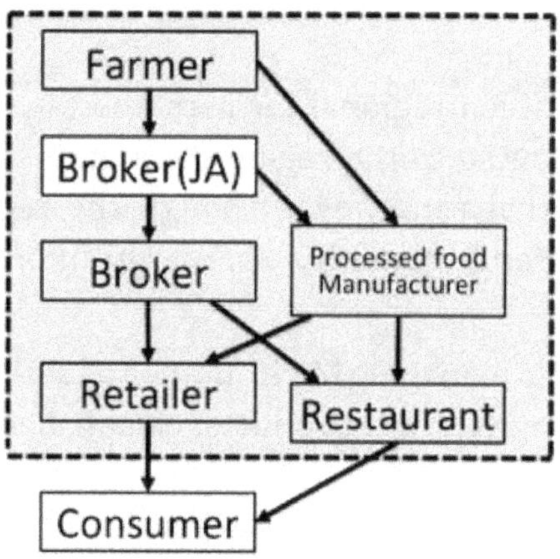

Figure 2. Integrated value chain by broker.

In these two cases, regional JA organization has very tight relationship with regional farmers, and they were involved in effortful marketing, direct retail service, and also restaurant service. Their businesses were also covering processed food manufacturing. The value chain of agriculture was completely reformed and integrated as schematically indicated in Figure 2, and as a result, it gained the profit of farmers.

4. CASES OF DIVERSIFICATION BY PROCESSED FOOD MANUFACTURERS

4.1. Mr. Fumiya Hamamachi and Ice-Cream Business Utilizing Regional Agricultural Products

Mr. Fumiya Hamamachi was originally a fisherman in the rural village. He was engaged in fishery when he was a teenager. However, as he had a dangerous experience in the sea, he quit fishery and changed his job to sales, by working for a certain enterprise. He was involved in sales and marketing of ice-cream items that his enterprise purchased from a certain manufacturer. Spending many years of much effort, he got some stable customers for ice-cream items. But the enterprise changed their strategy of sales and gave up selling ice-cream items. This was the motivation why Mr. Hamada began to establish his own business to sell ice-cream items. Because he had not sufficient capitalization at that time, he should have made much effort to establish his own business. During the hardship of establishing his business, many village people helped him by their primitive personalities and kindness. Those experiences affected him and produced the strong feeling of appreciation to region and regional community, he mentioned later. Gradually, he had advanced his business and made up his mind to initiate the business of manufacturing ice-cream products by his own factory [5].

One remarkable characteristics of Mr. Hamada was his excellent capability to find out delicious fruits/vegetables and to make a friend with the regional farmers who produce such delicious products. Because he loves region and regional community so much that the farmers were willingly accepting special contracts with him about providing their delicious products. As a result, Mr. Hamada's ice-cream products utilizing many delicious fruits/vegetables became popular and famous even for consumers in urban area.

Mr. Hamada's another excellent capability was on marketing and promotion. Because of his long-time experience/effort on sales, his capability on communicating with consumers and retailers became excellent. He always visited the retail front by himself whether it is domestic or abroad and communicated with people directly.

Again, in this case, Mr. Makoto Umebara, a charismatic designer, played an important role to promote the enterprise of Mr. Hamada and his products. Mr. Hamada asked Mr. Umebara to design the package of Ice-cream items and make a catchy for his enterprise. "Running about for seeking a delicious food" was the catchy that Mr. Umebara came up with.

In this case, Mr. Hamada played a role of catalysis and bridge between farmers and market. Usually, regional farmers are not good at communicating with consumers in market because they are concentrated only in cultivating like a craft person. So that, the information and the advice on the market, that Mr. Hamada could give them, were so important and effective for the farmers. They got the direction of products and became proud of themselves by answering to the request from Mr. Hamada. Of course, Mr. Hamada's enterprise is independent from farmers, but actually the mission of the enterprise is contributing to region and regional farmers; therefore, the activities of Mr. Hamada enlarged the business of the farmers and integrated the value chain of agriculture.

4.2. Hamada Family and their Cake Business Collaborating with Local Farms

Hamakou Corporation is a famous cake manufacturer, owning shops in local city, Kochi, and was founded in 1952. The founder of Hamakou Corporation was a professor of a local university. He was very curious about western cakes particularly Baumkuchen, a round-shape cake developed in German. He personally researched how to make such a cake, and after spending many years with much effort, he successfully realized the cake by his own cocking manner. This is an origin of the famous cake manufacturer named Hamakou. Now, they have over 200 employees and annual sales of around 1.7 billion yen (around 17 million US dollar), and their business is covering processed food manufacturing, managing 16 shops and a hotel including restaurants and also a fruit farm. The current CEO, Mr. Yukihiro Hamada, is a ground son of the founder.

As a manufacturer of processed food, they have been developed a variety of items such as Western cakes, Japanese cakes, jams, beverages, jelly, local beer, and so forth. Their policy was to think their business as cultural activities. That is, to make a cake was a kind of transferring the regional culture, they thought. Particularly, Japanese cake was an enjoyment on beauty of change of seasons, which was one of the remarkable characteristics of Japanese nature, they mentioned. They emphasized that Japanese people historically have been enjoying Japanese nature and its beauty by tasting traditional cakes. Also, they respected special agricultural products in region and insisted to use those products for their cakes to convey what a wonderful treasure the region has to regional people. They also mentioned that the founders' son, the second CEO, has learned many things from French people. Particularly, one thing was that they respected their regional agricultural products when they make jerry cakes. These thinking ways and policies were transferred as a family motto through family business of Hamakou.

Based on the motto, they also initiated hotel business over 20 years ago. On the beautiful hill facing seashore, they have been managing a hotel with a restaurant and a fruit farm by themselves. The purpose of this business is direct transferring the attraction of regional fruits and vegetables to customers. They are also involved in producing processed food products as OEM (Original Equipment Manufacturing) for many other regions' enterprises. The purpose of this activity came from their mission as to transfer the charms and attractions of the regional agricultural products to many people in other regions. Their products are the tools for communication with people and for sharing joy of regional culture/nature, they mentioned.

They themselves design the packages of their products and develop new items because they mentioned that they are the best persons to know their motto on their items. Developing new items is one of the most important subjects of their business. They developed a famous item named "Kanzashi (meaning "ornamental hairpin")," which marked a big hit called as one of the masterpieces representing regional cakes. After the break of this item, they have been continuing the effort to improve by testing a variety of devices over 200 hundred times for 4 years. They always continue improvement of products by absorbing customers' voices and demands.

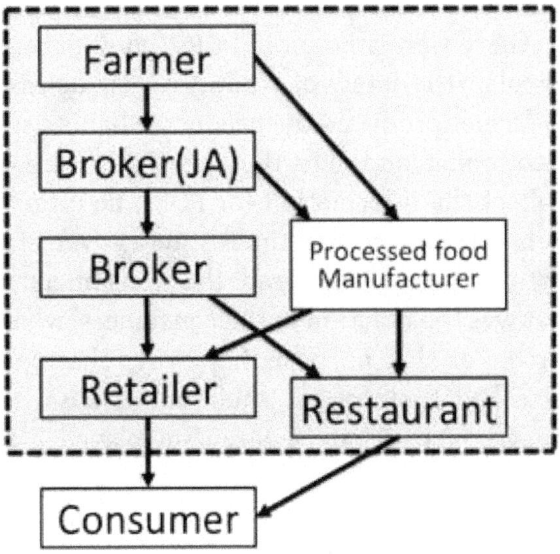

Figure 3. Integration of value chain by manufacturer.

To realize such a high-quality service, they respect communication among employees. Everyday, the CEO reads whole members' daily reports and informs them his thinking at least once a week. In order to share the affection on cakes not only with customers but also with every employee, daily communication is very important, the CEO mentioned.

Their products were selected as the second excellent item in all of the regional food products in nation by Japan Air Line (JAL) and were served for business class seat customers in the airplane of JAL.

The activities of Kochi Ice and Hamakou are schematically indicated in Figure 3. The CEOs of these enterprises played a role of binding market and regional farms. Mostly, the businesses of processed food manufacturer are designed business carefully to match the market and the preference of consumers because those people are located much closer to market than framers. So that manufacturer of processed food can be a good guide of market for framers. If regional manufacturer has a strong hometown feeling and solidarity with regional community, they could establish tight relationship with regional farmers and could contribute to integrate the agricultural value chain.

5. CASES OF DIVERSIFICATION BY RETAILERS

5.1. Mr. Kouji Usui and his Activities on Vegetables Business

Mr. Kouji Usui was working for a certain greengrocery when he was young. The name of the green grocery was Tosa Senri. Tosa was a name of the place of the farmers who supply the vegetables to this grocery, and Senri was the name of the place where the grocery was located and the consumers' living area. As indicated in the name of the shop, the owner's policy was bridging consumers and framers. So that, in this grocery, there were amount of POPs (point of purchase advertisings), which describe and explain the details of product items, such as where is the place of production, who is a farmer produced the product, what is a special characteristics of the product, how to cook it, and so forth. Those POPs were written by Mr. Usui's hand directly. To collect the information for POPs, he often visited the farmers' places and asked them directly by sometimes staying over in the farmers' houses. Also, at that opportunity, he transferred the consumers' information to the framers, such as what was the reaction of the consumers, what they said about the products, and so forth. For that purpose, he always chatted with the customers in the grocery and asked them about their opinion, satisfaction, impression, and request for the products. That is, Mr. Usui played a role of exchanging/ transferring information and bridging between farmers and consumers. On the surface, the business of this grocery was selling fruits and vegetables. However, their business had an aspect of information service. Their manner of transferring information depended not on information technology like Internet but on human power, somewhat old-fashioned style method. However, it was so good to transfer the detailed information from human to human, rather than the transferring by Internet or some other IT methods. By using human network transferring, their information was very deep, detailed, and sensitive.

At that time, the annual sales of the grocery were so good and reached up to over 1 billion yen (around 1 million US dollar) by only two employees, Ms. Mariko Tomita, leader, and Mr. Usui, sales staff. Such high sales were very unusual in comparison with most of the groceries dealing with regional products. Generally, it was very difficult for ordinary groceries to gain such high sales. One reason of their success was that they insisted to provide only excellent products that they themselves could be satisfied with the quality. Another was that they transfer the information on demand and supply between customers and farmers to create a new product fitting to consumers' preference. By the effort of Mr. Usui, farmers and consumers both got benefits on production and consumption. In this case, the retailer played a role of binding both sides. After working for Tosa Senri, Mr. Usui initiated his own business focusing making POPs for many shops because he recognized that his capability on information transfer would be useful for many retailers.

5.2. Mr. Masaharu Tamamori and Iriomote Pineapple

Mr. Masaharu Tamamori was a vice president of the tourism enterprise named Yaeyama Tourism Cooperation. The main business of the enterprise was operating a ferryboat among many islands called Yaeyama islands. But they also managed many other businesses such as operating shuttle bus, taxi, hotels, super market, and so forth in region. Because everyone should collaborate with each other in an island, a small community, the enterprise also should be involved in many kinds of businesses regarding the daily lives of islanders, including cultural events such as summer festival, collaborative activities of cleaning public spaces, and so forth.

In the past, the main industries of Iriomote island were growing sugarcane and pineapple. The farmers sold pineapples to the processed food factory at the price of only 25 yen in the past, they mentioned. However, by globalization, the lower price pineapple made in Taiwan and Vietnam surpassed their products in the market, and the factory of Japanese pineapple was closed. The farmers on pineapple lost their job as a result.

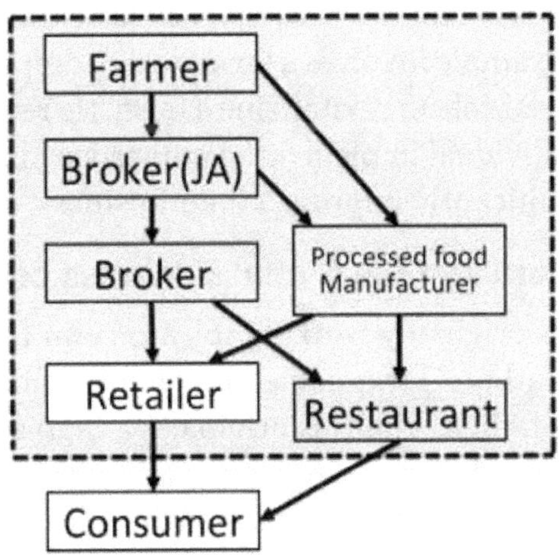

Figure 4. Integration of value chain by retailer.

Mr. Tamamori, a younger member of the board of trade of the island at that time, watched the distress of the farmers and made up mind to support them by selling their pineapple. He ran about all over nation to promote the pineapples of Iriomote. He went to many railway stations and sang a traditional song of Iriomote with ethnic instrument to appeal that Iriomote was a peaceful island with beautiful nature. After over 10 years' effort of promotion, Iriomote pineapple got popularity among Japanese people and many sight-seeing tourists tended to purchase it with a higher price of several hundred yen, over 10 times that for the factory in the past.

Nowadays, the proliferation of Internet and delivery system accelerated the business of Iriomote pineapple, so that Iriomote pineapple became one of the famous souvenirs of Yaeyama islands.

In this case, the relationship between farmers and the key persons was very tight. In the small community like an island, every one should collaborate with each other for survival. The binding among islanders was so strong that everyone has a strong passion for contributing to island based on their solidarity. In that case, if the key person has a capability of business management, he/she can contribute to build an effective integration of agriculture value chain.

The two cases described here have the same characteristics, tight binding between farmers and retailers and also close communication between farmers and consumers through the medium of retailers. The realized integration of food industry is schematically indicated in Figure 4.

6. CASES OF DIVERSIFICATION BY A FARMER WITH BUSINESS EXPERIENCE

Several situations discussed in this chapter had crucial individuals who were not farmers. Typically, farmers specialize in growing vegetables or fruits and may lack expertise in business or sales. Primarily, they lack knowledge and expertise in business or sales. This example involves a farmer who, despite having a background in business, plays a crucial role in revitalizing a farm. He restructured the farm into a firm. The essential individual implemented numerous modernizing technologies on the farm, which significantly improved their business.

6.1. Mr. Yuji Nakamura and his Global Business on Lily Bulb

Mr. Yuji Nakamura was originally working for a certain life insurance enterprise before entering to his fathers' farm named Nakamura Farm. Nakamura Farm was founded in 1955 by his father, who was originally a high school teacher majoring biology. His father was much keen on flowers, particularly on lily. He launched his business of producing lily bulb after long time researching growing technology of lily bulb.

In 1990, Mr. Yuji Nakamura succeeded his fathers' business and established an enterprise organizing Nakamura Farm because of bad health condition of his father. He initiated global business on lily bulb by utilizing his capability and knowledge on modernized business.

One excellent idea was that they initiated lily bulb business in winter by carrying bulbs by refrigerating container from other countries in the Southern Hemi sphere such as Chili, New Zealand, and so forth. Usually lily cannot be grown in cold circumstance like winter, so that there was no supply of lily bulb in winter. But they grew lily bulbs in hot circumstance in the Southern Hemi sphere and brought them to Japan by using refrigerating container and defrosted them by their original technology. By this device, they could occupy winter market of lily bulb in Japan.

They positively learned many things from Netherlands with advanced technology on flowers and imported lily bulbs and some other flowers' bulbs from abroad. They also built greenhouses controlled by computer to research the best growing conditions for lily flowers. They sold lily bulbs with precious information/ data on how to grow lily flowers, obtained through their research in greenhouses. It was a good service for customer farmers who would grow lily flowers by purchasing lily bulbs from them in order to sell the grown flowers to flower market.

Currently, their annual sales reached over 1.7 billion yen (around 17 million US dollar) by only 16 employees. Nakamura Farm is so popular for job-hunting students that over one hundred students applied for only one opportunity of recruit.

Nakamura Farm is also highly respecting relationship with regional community because their business and daily lives are strongly bound with regional community. They held a follower festival once a year to express appreciation for regional community. They have also many social activities collaborating with regional schools.

In this case, the key person in the side of farming had strong passion to contribute to regional community, and he had also capability of modernized business management. This is somewhat a rare case in current status in Japan, but the number of the persons like Mr. Nakamura is getting increased gradually. The effort of agricultural diversification will be much more fruitful in the future.

7. ANALYSIS ON KFS AND DISCUSSION

To analyze Key Factor of Success (KFS) of the cases presented in this chapter, the common elements in the cases particularly focused on character/personality of key persons are summarized as follows.

- In every case, the key person had a strong entrepreneurship and capability of business management. Usually the farmers are not good at business itself because of their specialty and mentality. In that sense, the key persons who are good at business management can contribute to actual integration of value chain or re-allocation of farmers' assets/resources in new business style.
- In every case, the excellent strategies of advertisement, marketing promotion, and branding are common elements for inducing their business success. Because most of the poor regions are lack of resources, branding is an effective manner to lift up their business.
- In the most of the cases, the market size of their products was not so large that large competitive enterprises will not penetrate the market to enlarge their business. If the market is so large, cost-reduction, and price-down by scale-merit of large enterprises with large capitalization will be a threat in business.

- In every case, the key person had a strong feeling of contributing to the regional community. Particularly, the case of Kochi Ice indicates that the kindness of local people made an important role for forming the personality and the business policy of the key person. Social capital as solidarity, trust, credibility, and hometown feeling in region is said to be essential to grow entrepreneurs with strong mission for contributing regional community [6, 7, 8, 9, 10, 11]. Because agriculture industry is commonly strongly linked to regional community, improving social capital in region is significant to advance agriculture by diversification.

The common factors mentioned in 1–3 are understandable by the following tendency. That is, generally, there are two directions of construction of industry as follows.

- Dividing value chain and horizontal collaboration for enlarging scale of industry for cost reduction
- Binding value chain and vertical integration for enhancing brand value in niche market

Emancipation of farmland and forming agricultural cooperatives by the policies right after the Second World War were executed for the purpose of activating degraded food industries in Japan, so that the policies were thought to be based on the above-mentioned direction a.

In contrast, the policy of agricultural diversification is proposed for the purpose of increasing profit of farmers by branding in niche market, so that the policy is rather thought to be based on the above-mentioned direction b. The common elements mentioned in 1–3 could be understood in this aspect.

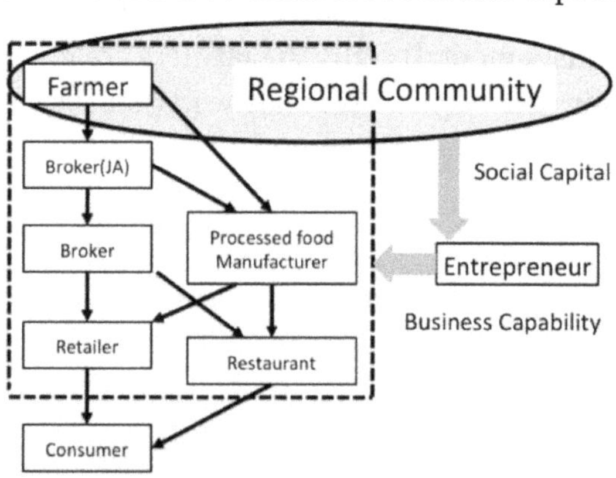

Figure 5. Circulation of intangible asset in region.

One of the implications induced from the case analysis is that successful key persons for agricultural diversification have strong entrepreneurship, capability of business management, passion and mission for contributing to regional community, and tight relationship with regional community. Although MAFF promoted the policy of diversification, the reforming social system to support the policy is not enough for realizing the policy. The activities of those kinds of key persons are essential in the real reforming of industries. The significance of human resource should be re-considered for the future of agriculture.

The shared components referenced in 4 are believed to suggest a form of social capital circulation in the region as an intangible asset. Social capital in the region fosters passionate entrepreneurs dedicated to contributing to the local community. These entrepreneurs drive business success by integrating the value chain, leading to profits from agricultural diversification that benefit farmers in the region.

This part presented numerous successful cases, but it also acknowledged the presence of failures in the actual industry. An aggressive entrepreneur in the food market may struggle to develop an efficient value chain integration without collaborating with farmers. In all successful instances, a strong connection between farmers and entrepreneurs was crucial for creating a business strategy. Social capital between regional communities and entrepreneurs can be seen as a valuable asset for advancing business and society. Asset circulation, as shown in Figure 5, will lead to an expansion in the industry, resulting in potential growth of asset/equity in society.

Another consequence derived is that the circulation of intangible assets plays a crucial role in achieving sustainable regional development. Not all types of agricultural lands will be suitable for implementing the companies discussed in this chapter. The agricultural diversification plan would be effective in specific specialized markets. The circulation of intangible assets in these enterprises could be a key factor in achieving sustainable development in resource-poor countries.

8. CONCISE SUMMARY

Japan has advanced agricultural diversification to address low profitability and market demand mismatch in the agriculture value chain. The Ministry of Agriculture, Forestry, and Fishery (MAFF) in Japan introduced the policy of agricultural diversification. However, successful implementation of the policy requires the cooperation and effort of various stakeholders beyond just farmers. This is because traditional farmers may lack the ability to coordinate other stages of the value chain.

Aggressive critical individuals like as brokers, processed food producers, and retailers play a crucial role in supporting profitable agriculture by partnering with local farmers. This chapter showed a total of seven successful cases of revamping agricultural value chains. Successful businesses focused on agricultural diversification were established by individuals with great entrepreneurship, business acumen, and a passion for contributing to the area community. The case analysis suggests that the actions of key individuals are crucial for the actual transformation of industries. The importance of human resources in agriculture's future needs to be reassessed.

Another conclusion is that the social capital of an area is crucial for fostering key individuals with entrepreneurial skills, fostering collaboration with farmers, and ensuring that earnings are reinvested back into the farming community and the region. Recognizing and promoting the circulation of intangible assets like social capital in a region is essential for the sustainable development of underprivileged areas. Agricultural diversification policies may not be universally effective across all sectors of the agriculture industry. The idea would enhance the profitability of farmers in specific specialist markets but may not be as effective in larger mainstream markets with economies of scale. The policy would serve as a supplement to promote modern Japanese agriculture.

9. AT A GLANCE

Japanese agriculture has seen various challenges in the past few decades, including low profitability, labor shortages, farmland abandonment, and decreased competitiveness against low-priced imported goods. One government initiative to promote Japanese agriculture is agricultural diversification, a strategy that involves rearranging the value chain to increase farmers' profitability. The strategy involves integrating various elements of the food industry value chain, such as agricultural production, processed food manufacturing, and food retail and restaurants. By reallocating farmers' business resources, farmers can become more active and increase their profits. For many years, the Japanese agricultural supply chain has heavily relied on Japan Agricultural Cooperatives (JA), leading to problems including demand-supply mismatch and poor profitability for farmers. The agricultural diversification policy was suggested to encourage the integration of new value chains and the restructuring of supply chains to address these problems. This chapter showcases successful instances of agricultural diversification in Japan and identifies the Key Factor of Success (KFS) for these endeavors.

REFERENCES

1. Cook ML. The future of U.S. agricultural cooperatives: A neo-institutional approach. American Agricultural Economics. 1995;77(5):1153-1159
2. Koyama R, Kobayashi K. Characteristics of Japanese Agricultural Cooperatives with special reference to institution and members' attitudes. Journal of Commerce, Economics and Economic History. 2007;75(4):33-48

3. Agricultural Cooperative Division, Agricultural and Forestry Economic Bureau Ministry of Agriculture and Forestry. Statistical Abstracts of Agricultural Cooperatives Japan. Japan: Agricultural and Forestry Economic Bureau Ministry of Agriculture and Forestry Japan; 1958
4. Hirano M. Regional Development through Ecological Businesses: Unique Cases in Japanese Rural Regions. London: Routledge; 2013
5. Hirano M. A role of entrepreneur for innovating regional agriculture: Through a case of Japanese Ice-Cream firm. In: The 17th Uddevalla Symposium; Uddevalla; 2014
6. Bauernschuster S, Falck O, Heblich S. Social capital access and entrepreneurship. Journal of Economic Behavior & Organization. 2010;76:821-833
7. Davidsson P, Honing B. The role of social and human capital among nascent entrepreneurs. Journal of Business Venturing. 2003;18:301-331
8. Doh S, Zolnik EJ. Social capital and entrepreneurship: An exploratory analysis. African Journal of Business Management. 2011;5(12):4961-4975
9. Liao J, Welsch H. Roles of social capital in venture creation: Key dimensions and research implications. Journal of Small Business Management. 2005;43(4):345-362
10. Westlund H. Multidimensional entrepreneurship: theoretical considerations and Swedish empirics. In: Proceedings of the 50th Anniversary Congress of the European Regional Science Association; August 2010; Sweden
11. Westlund H, Larsson JP, Olsson AR. Start-ups and local entrepreneurial social capital in the municipalities of Sweden. Regional Studies. 2014;48(6):974-994

06 DAIRY VALUE CHAIN IN VIETNAM: EVIDENCES FROM BAVI AREA

1. INTRODUCTION: AN OVERVIEW

Each research will interpret the value chain in various ways. Khoi [1, 2] states that "Chain" focuses on the vertical sequence of operations involved in the distribution, consumption, and maintenance of commodities and services. The chain consists of dynamic characters that are repeated in a specific order. The value chain concept was first established in 1985 by Porter in his book "Competitive Advantage: Creating and Sustaining Superior Performance." The value chain is defined in research as the concept of viewing a manufacturing or service organization as a system composed of subsystems with inputs, transformation processes, and outputs. Inputs, transformation processes, and outputs encompass the procurement and utilization of resources such as money, people, materials, equipment, buildings, land, administration, and management. The way in which value chain activities are executed impacts costs and influences earnings. Porter delineates a fundamental value chain of nine stages, categorized into primary operations within the chain and supporting activities. The initial key task in the inputs supply chain involves receiving and storing raw materials for an industry or specific sector. The production activity is directly linked to developing the highest value-added products by processing raw materials into the end product. The logistics department will handle the reception, storage, and distribution of the final products to dealers and retailers. Marketing and promotional media are used to promote the goods to consumers. Service or after-sales activity pertains to customer care aimed at retaining customers and increasing the product's value. Supporting activities, albeit not directly responsible for creating value in products, play a crucial role in enhancing the value of products and supporting the key activities that contribute to the value creation in every stage of the chain. The notion of the value chain in the dairy milk sector involves a series of actions that encompass the journey of dairy milk from producers to consumers. The combination comprises the farmer, responsible for raising dairy cows and collecting milk, the processors, and the sale. These actors are actively involved in the value chain and their interactions are reliant on the movement of goods, information, and financial transactions. Furthermore, the operations of this value chain were impacted by various elements like the supply chain system, marketing strategies, legislative framework, and the supply and demand of goods. Thus, a standard dairy milk value chain includes production, transportation, chilling and bulking, processing, transportation or distribution, and retail.

Dairy farming in Vietnam has been present since the early twentieth century, benefiting from the country's advantageous natural conditions as an agricultural nation. Since then, Vietnam's dairy industry has continuously evolved and made a substantial contribution to meeting food requirements. In Vietnam, the overall demand for milk rose steadily, with a specific 61% growth in the demand for raw milk. In 2010, fresh milk output was 306.7 thousand tons, and it climbed by 15.6% to 732.2 thousand tons. Yet, Vietnam's dairy sector is unable to meet the country's milk consumption needs. Milk retail prices in Vietnam are exorbitant, but dairy farmers sell fresh milk at significantly lower prices. This paradox stems from the overwhelming control of dairy companies in both the quantity and quality of the milk input. The process led to an unequal distribution of earnings and benefits among all participants in the dairy value chain. This chapter will analyze the allocation of benefits, costs, and value-added among the stakeholders in the dairy value chain in Vietnam, while also identifying issues within the management system.

2. DAIRY VALUE CHAIN IN VIETNAM: INSIGHTS FROM THE BAVI CASE STUDY

2.1. Literature Review

The literature is abundant with the works that address the value chain of dairy milk. These researches pointed out the important results in developing the value chain of dairy milk in general and each actor in chain. Thanks to Porter [3] and Kaplinsky and Morris [4], the concept of "value chain" and the method to calculate value-added and profit of actors in the dairy value chain are created. Lowe and Giraffe (2009) showed a good view of the dairy value chain in the USA—an advanced country in milk production. The value chain includes four main actors: inputs, production, process and distribution, and marketing. In the USA, high technology is applied in most parts. There are concentrated steps in nurturing, harvesting, collecting and processing milk. Every step is carefully controlled and managed. The special thing in the US dairy industry lies in the veterinary system. The result of the study is that the companies downstream from the dairy producer category, milk and dairy processors, include large, diversified companies despite being diversified well beyond milk and dairy products, nonetheless include companies that have higher shares of the dairy market than the largest producer cooperative. In the "dairy value chain in Kenya" (August 2008), a report by Techno Serve Kenya, East Africa Dairy Development mentioned each of the actor's situation in the dairy value chain in that country. In addition, the study also referred to the logistics operations in parallel in the series. By means of descriptive statistics, the study showed the limitations and the dominant presence of the value chain. Research also offered solutions and proposals to overcome the difficulties. Achchuthan and Kajananthan (2012) found out the main factors that have influence on the dairy sector of Sri Lanka. This study also discovered the strengths and weaknesses of each actor in the dairy value chain; discovered the strong, weakness, opportunities and even threats of each actor; to suggest actors in the dairy value chain to strengthening the dairy sector. About the strengths, the author pointed out that the farmers are being provided the financial assistance by some local companies to widen their farms and increase the scale of dairy milk production. The farmers are also trained about how to use modern equipment to test the quality of milk and educated about marketing strategy to develop their milk brand. Also, in Sri Lanka's agriculture sector, there is always abundant labors with low labor cost to expand their business. On the other side, the dairy farmers also have many difficulties in their business such as they do not have the educational background to plan the dairy farming in the large scale; further cooperative society in the Karachi division has not enough technological facilities to preserve the pure milk. And they also do not have the value-added strategies like milk toffee, ice cream, yoghurt in the large scale, etc.

Besides, in Vietnam, Tran Huy Cuong and Bui Thi Nga (2012) analyzed the actors in the value chain of the fresh milk in Vietnam. This research used not only quantitative but also qualitative approaches in the case of fresh milk products in the northern area of Vietnam. The research illustrated the four main actors in the dairy milk value chain in Vietnam: (1) the farmer; (2) collector; (3) processing firms and (4) distribution. The dairy plants play an important role and have right to make decisions. In other word, the dairy plants (processing firms) become the main actor, which receives most of the profit. The farmers who invest lots of capital and time to raise cow only receive a small proportion of the profit. This research gives a deeper vision on what is going on in the fresh milk production and distribution in Vietnam. This is one of the very few researches in Vietnam on the value chain of the dairy industry. Agreeing with the approach of some study, the study by Khoi [5] has provided quite detailed analysis on comprehensive value chain of Vietnam dairy industry to produce the value-added level in the value chain actors of Vietnam's dairy. This is a reference to help paper to get different perspectives on the development of value chains in different sectors in Vietnam. In another study of Vietnam by Khoi [6], the author gave the merger may occur in the dairy sector in Vietnam due to differences in value-added in the value chain of this sector.

2.2. Methodology

A survey was carried out in Bavi, Hanoi, where a large amount of fresh milk is produced annually. The content of the survey was built before conducting in Bavi with 160 questions. The questionnaires were based on the criteria which reflect the main objectives of the paper such as the actor's profits and costs, the relationship among these actors, and the advantages and disadvantages in the dairy milk's value chain. Some part of results of the paper should be evaluated (by interview) of the relevant target groups in the value chain of dairy milk.

Data collected were aggregated and analyzed by SPSS20, Excel. The calculation of cost and profit margin of each actor in the chain will also be presented by a quantitative tool for value chain analysis.

2.3. Overview of the Dairy Sector in Vietnam

In the last 10 year, Vietnam has witnessed the gradual growth of the milk consumption. Vietnam is not the country with long history of milk production; however, in the last few years, Vietnamese people have realized the importance of milk products and spend more on this nutria drink. The growth rate is quite high on average; nevertheless, Vietnam is still in the low milk consumption area (<30 kg/capita/year). This might be a big chance for milk producer in the near future.

Most milk products in Vietnam are designed for under 3-year-old babies and the elder. There is a huge market for daily milk consumption and mature specialized milk. According to Khoi [1, 2] 10% of Vietnamese population in the two big cities, namely Ho Chi Minh City and Hanoi, has been consuming 78% of dairy products. These data show an inequality in the consumption of milk products.

2.3.1. Dairy Milk Consumption

Gerosa and Skoet (2012) investigated 144 countries about the consumption of goods. They released an elasticity to show the range of demand. The income elasticity of expenditure estimates the percentage increase in expenditure on the food category resulting from a 1% increase in income. According to Table 1, we have the elasticity of the dairy milk product.

The income elasticity of expenditure estimates the percentage increase in expenditure on the food category resulting from a 1% increase in income. The numbers reported are simple unweighted averages of estimates for the individual countries included in each income group. Vietnam—a lower middle-income countries have the relatively high dairy elasticity 0.79, which might be a reason for the fast growth of the milk industry in Vietnam.

Table 1.The elasticity of consuming products.

	Low-income countries N = 28	Lower middle-income countries N = 36	Middle-income countries N = 36	High-income countries N = 44
Food beverages and tobacco	0.81	0.77	0.70	0.54
Beverages and tobacco	1.73	1.13	0.92	0.67
Cereals	0.59	0.49	0.34	0.08
Meat	0.80	0.76	0.69	0.53
Dairy	0.83	0.79	0.72	0.55
Fish	0.69	0.64	0.56	0.42
Fats, oils	0.60	0.50	0.37	0.15
Fruits	0.66	0.60	0.51	0.36
Other foods	1.82	1.23	0.98	0.70

The practical situation in dairy milk industry in Vietnam was unacceptable with Gerosa and Skoet's research (2012). Along with the increase in the living standard, the amount of annual consumption of fresh milk in Vietnam increased over the years, especially in the period from 2010 to 2015. In the 10 years since 2000, the number of annual milk consumption in Vietnam increased only 2.8 kg. However, in the period 2010–2015, the annual milk consumption is 5.8 kg, which is more than double amount of 2010. In the period from 2010 to 2013, the amount of milk consumed per capita per year increased only slightly due to the affecting of the melamine crisis in 2010. After recovering, from 2013 to 2015, Vietnam has witnessed the significant increase from 4.0 to 9.3 kg/capita/year (Figure 1).

2.3.2. The Dairy Milk Production

The number of dairy cow and quantity of milk production in Vietnam is increased rapidly from 2000 to 2010. From 2000 to 2004, the number of dairy cow increases steadily from 35 thousand to nearly 100 thousand heads. The reason for that is the increase in the demand of milk and the implementation of resolution 09 and decision 167, which created condition and opportunity for Vietnam's development by stimulating Vietnamese farmers to raise cows (the resolution contributed to increase the dairy herd to 100,000 milking cows and production of 300,000 tons fresh milk yearly). The decision built a production and development strategy for dairy cattle for the period 2002–2010. This strategy aimed to raise dairy cow in Vietnam including 10 provinces in the North, 5 provinces in the Central Coastal region, 4 provinces in the Central Highlands and 7 provinces in the South. In 2005 and 2006, the growth rate of the dairy cow number was only 8.7% per year; which resulted from the inefficient milk production due to the low technology. However, with the restructure of dairy program, accompany with the increasing quickly in milk demand. The number of dairy cow recovered in raised 14.9% per year from 2008 to 2010.

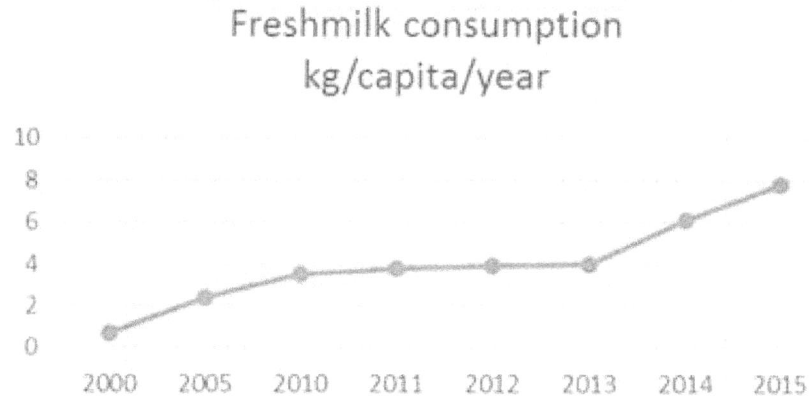

Figure 1. Total fresh milk consumption in Vietnam from 2000 to 2015. Source: General statistic office of Vietnam.

From 2010 to 2015, the number of dairy cows increased gradually with the average of 11.73%. In 2014, Vietnam has 200,400 dairy cows. As a result, total dairy milk production in Vietnam in 2010 was 306,620 tons; it was six times higher than the dairy milk production in 2000. From 2010 to now, the quantity of milk production has continuously increased. In 2015, Vietnam has witnessed an increase in fresh milk production by 15.6% compared with 2010. The quantity of fresh milk in Vietnam in 2015 was 723.2 thousand tons (Figure 2).

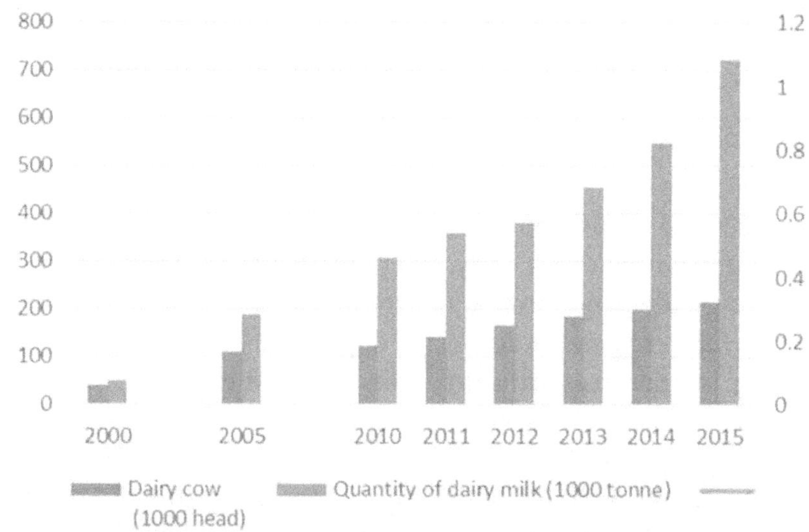

Figure 2. Quantity of dairy cow and milk production in Vietnam from 2000 to 2015. Source: General Statistics Office of Vietnam.

2.3.3. The Imported Quantity of Milk

Due to the melamine milk crisis in the Asia region in 2008 and 2009, the number of imported milk of Vietnam had decreased suddenly in early 2010. From then to now, the quantity of milk that Vietnam imported has recovered gradually. Currently, in the end of February 2016, Vietnam's import reached to 168 items, 8 million of dairy products and increased 17.99% with last year. Vietnam imported raw milk and other milk products from 16 countries around the world, mainly from New Zealand, accounting for 35.1% of total turnover, reaching 59.3 million, and raised 41.58%. The second largest supplier is Singapore reaching 22.4 million, and increased 8.4%; followed by Australia, reached 16.7 million (data 2016) Generally, in the first 2 months of 2016, the amount of imported milk from other markets had positive growth rates such as Denmark, France, the USA, Korea, Philippine, etc.

2.4. The Dairy Value Chain in Vietnam

The value chain of dairy milk in Vietnam includes a lot of activities such as input supply; farming, production; bulk and cooling processing; packaging; transportation and distribution, which divided in five stages. There are many actors in the chain: A, feed, heifers machines; B, dairy men (dairy farmers); C, milk collectors; D, dairy plant; E, wholesaler, shop agents, showroom-market; F, Retailers; G, middle men (small shop, milk care shop); and H, Consumers (domestic consumers and international companies). Besides, the value chain of dairy milk was supported by other organization through projects and policies (Figure 3).

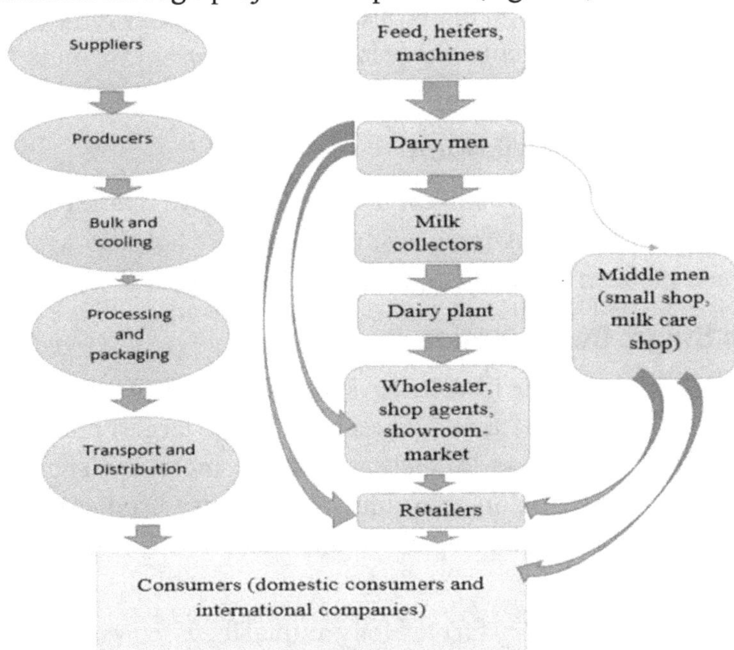

Figure 3. The value chain of dairy milk in Vietnam—the function and actors.

2.4.1. First Stage: Input Supply

Dairy cows and industrial food for agriculture sector derive from domestic market and importing from the foreign. However, the importing dairy cows and food accounted for a relatively large amount. The dependence on this importing source may effect on the stability of Vietnam dairy milk. Therefore, in stage of input materials, one of the most essential tasks of Vietnam dairy milk is being prior to choose the best dairy cows.

2.4.2. Second Stage: Production and Farming

This stage plays the most important role in the whole value chain of Vietnam dairy milk and it also the stage, which Vietnam still, have the huge gap with other countries. There are several reasons why Vietnam is not good at this stage. Firstly, it comes from the lack of animal food, infrastructure, technology and the support from Vietnam government. Secondly, the famers in the Vietnam's dairy value chain also lack the necessary knowledge and skills of livestock sector (dairy farming). As a result, the dairy breeding sector only meets 25% of domestic demand.

2.4.3. Third Stage: Processing

This stage is the combination of milk collecting, processing and packing. Presently, there are three major participants taking part in milk collecting: milk procurement companies, cooperatives and privates. Although the milk collecting companies have increased the level of milk procurement through their collecting points, however, with the participation of private in collecting dairy milk makes the price of milk is not stable. Additionally, there is still above 20% of the milk that does not meet the quality standards.

2.4.4. Fourth Stage: Distribution

Recently, in Vietnam, the dairy milk is distributed in two main trends: traditional channel (distributors account for over 10% profit - wholesale agents - stores - consumers) and modern channel (supermarket - consumers)

2.4.5. Fifth Stage: Consumption

The large of milk consumption in Vietnam shows a positive picture when the milk is one of the most important goods and it accounts for the largest market. In recently, the level of milk consumption in Vietnam market increased significantly. Only 10% of the country's population consumes 78% of dairy products. This promotes Vietnam dairy branches to develop more and more to meet different consumer needs.

In the case of Bavi, the dairy farmers buy animal feeds, cows and machines from suppliers and generate milk by their own. Dairy farmer mainly sold their raw milk to collecting centers by 97%; they keep the rest at home as food resource or provide for retailers and shops by 1 and 2%, respectively. There are many retailers and milk shops situated near highway from Hanoi to Phu Tho and Vinh Phuc, where it is convenient for anyone to buy milk. The reason why there are still a considered amount of milk that was provided to retailers (Figure 4).

From dairy plant, most of milk products (96.7%) were distributed to the wholesaler, shop agents, showrooms, only 1.2% of those products to schools and companies and 2.1% of them to the retailers. From wholesaler, shop agents and showroom-market, these products were distributed to retailer (68%) before being sold to consumers and 32% of those products are shared out directly to the consumers.

2.5. The Cost, Profit and Value-Added

2.5.1. Economic Analysis of Dairy Farmers

The survey was carried out in Bavi district with the aim to examine the production cost of dairy farmers in Bavi. With the including of family worker in the cost production, the total average cost per 1 kg of milk is almost 9793.9 Vietnam dong (VND). This cost contains the cost of food, vaccine, electricity, water, labor and transportation as shown in Table 2. Farmers plant nappies grass, corn in their field and harvest for cows to diminish the cost of food. Despite of that saving, they still have to pay 1,300,000 VND per cow for industrial food, which accounts for 56.72% of total cost. However, the average price of 1 kg fresh milk is only 11,000 VND. It is clear that the profit of farmers is only 1206 VND/ kg milk while they generate almost 4511 VND value-added.

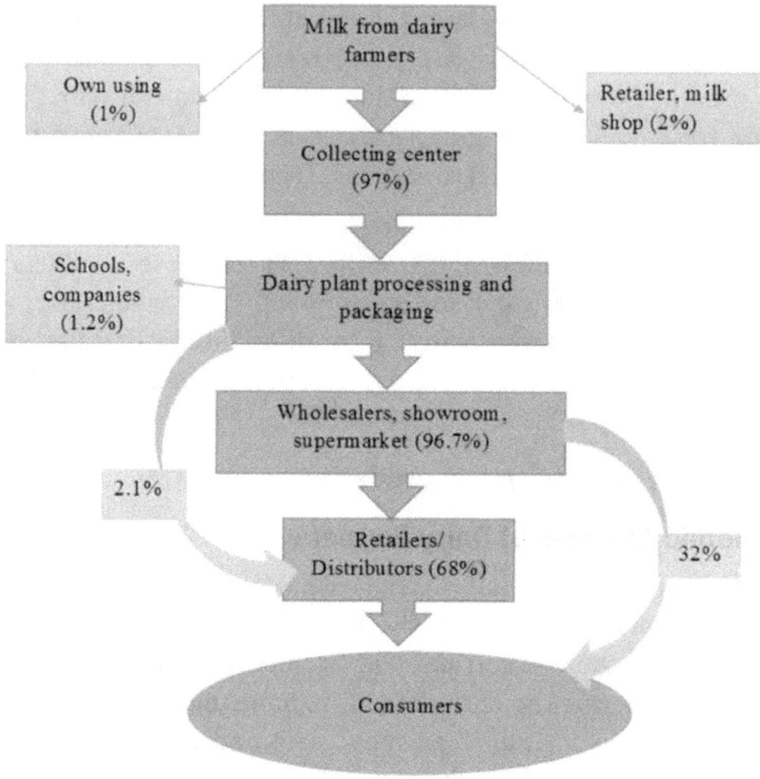

Figure 4. The distribution in the value chain of dairy milk in Bavi. Source: Calculated by data collected, 2016.

But in fact, when calculating the cost production of farmers, the labor cost of family members were not calculated. Thus, the intermediate cost was maintained, whereas the total cost decreased to 6987.6/kg milk. So that, with the price of 11,000 VND/kg milk, they believe that they could earn the profit at 4012 VND/kg of milk.

2.5.2. Economic Analysis of Dairy Plants

Firstly, this chapter will not calculate the cost, profit and value-added of the collectors in Bavi. The reason is that the dairy milk chain in Bavi is different from other value chain. In this value chain, the collectors belong to dairy plant; they are not households or traders so the economic analysis is added into the dairy plant's economic analysis (Table 3).

In general, 1 kg of fresh milk is 24,000 VND. Dairy companies usually have high intermediate cost, which accounts for 46.01% of milk price with 16,102 VND. Intermediate cost contains input material, which is imported from other countries, the input milk from dairy farmers, electricity to operate dairy plant, water and other materials. The average total cost of dairy plant is 18,423 VND (78.53%). Thus, dairy companies gain the average profit of 5577 VND, which account for over 23% of total revenue. Besides, they generate the amount of value-added at 7898 VND.

Table 2. Cost, profit and value-added of dairy farmers (including family workers).

	Items	Detail	Unit	A-Quantity	A-Price	A-Costs	A-Share
A	Cost						
	Food	Produce (grass, others)	VND/cow/month	6 cows	79,100	474,600	3.45
		Purchase	Kg	6 cows	1,300,000	7,800,000	56.72
	Vaccine		VND/cow/month	6 cows	20,000	120,000	0.87
	Electricity		VND/cow/month	6 cows	36,000	216,000	1.57
	Water		VND/cow/month	6 cows	50,000	300,000	2.18
	Labor	Family worker	Man month	2	2,000,000	4,000,000	29.09
		Hired worker (cutting grass)	Man month	4	160,000	640,000	4.65
	Transportation		VND/month			200,000	1.45
	IC (Income)		VND/month			9,110,600	
	Total cost		VND			13,750,600	100
	Total unit cost		VND/KG			9793.87	
B	Sale price		VND/kg	1		11,100	
	Revenue		VND/month	6 cows	2,574,000	15,444,000	
C	Profit		VND/cow/month	6 cows	175,500	1,693,400	
	Unit profit		VND/kg	1		1206.125	
D	Value-added		VND/kg			4510.97	

Source: Calculated by data collected, 2016.

2.5.3. Economic analysis of wholesalers

Wholesalers at level 1, who buy milk product directly for dairy companies; they are hired to sell product for companies. The advantage is that they do not need to buy facilities and the input products were not calculated in the intermediate cost. So that, all the input cost they have to pay is electricity and renting. The other cost is labor cost; the average wage of a worker or seller in the shop is 2,500,000 VND. In the end of month, the companies will assess the revenue of wholesaler and pay the commission for them. The commission is based on the level of revenue (revenue 10 million VND, get 10% commission; revenue 30 million VND, get 20% commission; and revenue over 50 million VND, get 30% commission). This method encourages the wholesalers trying to sell more and more.

On the other hand, the wholesalers at levels 2 and 3 (smaller scale) buy milk products of dairy companies in the low price and sell it to retailers and consumers with higher price to get profits. In this way, wholesalers have more independence in their businesses. They also were equipped facilities by companies (this is advertising and the taking market share of companies) (Tables 4 and 5).

Table 3. Cost, profit and value-added in 1 kg milk of dairy companies.

Items	Unit	Costs	Share (%)
GO/TR (Revenue)	VND/kg	24,000	
TC (Total cost)	VND/kg	18,423	52.64
IC (Income)	VND/kg	16,102	46.01
VA (Value-added)	VND/kg	7898	22.57
NPr (Net profit)	VND/kg	5,577	15.93
Source: Calculated by data collected, 2016.			

Table 4.Cost, profit and value-added of wholesaler at level 1.

Wholesaler 1				
Items	**Unit**	**A-Quantity**	**A-Price**	**A-Cost**
Cost				
Electricity	VND/month			2,100,000
Rent	VND/month			2,000,000
Labor	Man month	2	2,500,000	5,000,000
IC (Income) VND/month				4,100,000
TC (Total cost)	VND/month			9,100,000
GO/TR (Total revenue)	VND/month			55,000,000
Get	VND/month			16,500,000
NPr (Net profit)	VND/month			6900,000

Wholesaler 1				
Items	Unit	A-Quantity	A-Price	A-Cost
VA (Value-added)	VND/kg			2800
Source: Calculated by data collected, 2016.				

Table 5.Cost, profit and value-added of dairy farmers of wholesaler at level 2.

	Wholesalers 2 and 3			
Items	Unit	A-Quantity	A-Price	A-Cost
Cost				
Input	kg/month	1000	24,000	24,000,000
Electricity	VND/month			2,000,000
Rent and transportation	VND/month			2,000,000
Labor	Man month	1	400,000	400,000
IC (Income)	VND/month			28,000,000
TC (Total cost)	VND/month			28,400,000
GO/TR (Total revenue)	VND/month	1000	32,200	32,200,000
NPr (Net profit)	VND/month	900	4222.2	3,800,000
VA (Value-added)	VND/kg			3755.6
Source: Calculated by data collected, 2016.				

Finally, wholesalers could get an average profit at 6,900,000 VND and generate the value-added per milk product at 2800 VND. With wholesalers at levels 2 and 3, the average total revenue is 32,200,000 VND per month and get about 3800 VND/kg milk. They create the average value-added at 3755.6 VND, which is higher than that in wholesaler level 1.

2.5.4. Economic Analysis of Retailers

The retailers in the value chain of dairy milk in Bavi have the same business method with wholesalers at levels 2 and 3, but they have much smaller scale than wholesalers. Most retailers buy milk products at wholesaler and distribute to consumers. The average labor hired is smaller than one because they usually use their own labor force. The input price of retailers is 32,200 VND/kg milk and sale price is 35,000 VND. The average quantity milk sold per month is 200 kg. They could earn 1050 VND/kg milk and generate 1800 VND value-added (Table 6).

2.5.5. The Benefit and Value-Added Comparison Among Actors

Base on the method that Kaplinsky and Morris [4] has launched, the costs and profits are divided among the actors could be determined and thereby determine who may benefit from the chain.

In Table 7, it takes the cost A for the dairy farmers when they produce one unit of milk product; they sell milk product with price G. Follow the alongside of chain, H, I, J, K are price of one unit milk product of each actors. B, C, D, E are added cost when milk products transfer to the next steps including collectors, dairy plants, wholesalers and retailers. Hence, the total costs to produce one unit of milk product of each actor are: G + B, H + C and J + E. F = A + B + C + D + E is the total cost to produce one unit of milk product from the beginning stage (farmers) to the final stage of the value chain (consumers). So, the total profit that the chain get from producing one unit of milk product is F-K. From that, percentage of added cost, percentage of added-profit and percentage of retailer price of each actor in the dairy value chain will be calculated clearly. Based on Table 7 and the data about unit total cost and price of each actor mentioned above, the relative financial position of actors in value chain was showed in Table 8.

Table 6. Cost, profit and value-added of dairy farmers of retailers.

Items	Unit	A-Quantity	A-Price	A-Cost
Cost				
Input	kg/month	200	32,200	6440,000
Electricity	VND/month			200,000
Labor	Man month	0–1	150,000	150,000
IC (Income)	VND/month			6,640,000
TC (Total cost)	VND/month			6,790,000
GO/TR (Total revenue)	VND/month	200	35,000	7,000,000
NPr (Net profit)	VND/month	200	1050	210,000
VA (Value-added)	VND/kg			1800
Source: Calculated by data collected, 2016.				

Table 7.Relative financial position of actors in value chain.

Chain actors	Costs			Revenue	Profit		Margin	
	Unit total cost	Added unit cost	% Added cost	Unit price	Unit profit	% Total profit	Unit margin	% Retail price
Farmer	A	–	A/F	G	G-A	(G-A)/(K-F)	G	G/K
Collectors	G + B	B	B/F	H	H-B-G	(H-B-G)/(K-F)	H-G	(H-G)/K
Dairy plant	H + C	C	C/F	I	I-C-H	(I-C-H)/(K-F)	I-H	(I-H)/K
Wholesaler	I + D	D	D/F	J	J-D-I	(J-D-I)/(K-F)	J-I	(J-I)/K
Retailer	J + E	E	E/F	K	K-E-J	(K-E-J)/(K-F)	K-J	(K-J)/K

Chain actors	Costs			Revenue	Profit		Margin	
Total		F = A + B + C + D + E	100		K-F	100	K	100
Source: Kaplinsky and Morris [4].								

Table 8.Cost and margin for actors in Bavi in 2016.

Chain actor	Costs			Revenue	Profit		Margin	
	Unit total cost	Added unit cost	% Added cost	Unit price	Unit profit	% Total profit	Unit margin	% Retail price
Farmers	9793.9	—	41.91	11,000	1206.1	10.37	11,000	31.43
Dairy plant	18,423	7423	31.77	24,000	5577	47.94	13,000	37.14
Wholesalers	284,000	4400	18.83	32,200	3800	32.67	8200	23.43
Retailers	33,950	1750	7.49	35,000	1050	9.03	2800	8
Total		23,366.9	100.00		11,633.1		35,000	100
Source: Calculated by data collected, 2016.								

As presented on the data, it is evident that farmers incur high cost (41.91% of the total) and has very little profit (only account for 10.37%), whereas the wholesalers have little costs and relatively high profit with 32.67%. For the case of dairy plants (dairy companies), they added 31.77% of total cost—the second highest cost, however, it should be noted that, they get a remarkable profit with 47.95%. Overall, the financial position of actors indicated that the costs and margins are shared unequally in the chain. Specifically, for every 35,000 VND that a consumer pays for a 1 kg of milk product, 11,000 VND goes to farmers, 13,000 VND goes to dairy plant, 8200 VND goes to wholesaler and 2800 VND belong to retailers.

In the term of profit, the profit seems to be distributed unequally among actors along the chain of dairy milk. Farmers who invest most in their facilities, technology and food received only about 10% profit of the value chain. The profit receiving is inclined to the dairy companies, which get the highest profit with 5577 VND per one unit of milk product. Going to the end of the dairy milk chain, the profit decreases gradually with 32.67% for wholesalers and about 9% for retailers (Figure 5).

2.6. The Dairy Chain Management

Management activities of the dairy value chain in Bavi have many advantages such as:

- The value chain is concerned and encouraged as much as possible. Typically, farmers are supported in all aspects by the establishment of Bavi Cattle and Forage Research Center (BCFRC). To stimulate farmer in the increasing the number of dairy cows of the district, the BCAFRC collaborated with IDP companies (International Dairy Joint Stock Company established since 2004) to provide loans for farmers with 0% interest rate. With these loans, they could buy breeds and build cow house.

Up to 93% of farmers buy cows from the center and 7% of farmers buy cows from other household (small cow). Hundred percent of them implemented fully vaccinated policy and 100% of cow health managed by authorities.

- Control of the company's commitment: The implementation of milk consumption of dairy companies registered trademark. Bavi milk is always under the control, which ensures the fairness in trade by farmers.
- Milk quality management: At collecting stations of mainstream companies, there are always strict testing processes to ensure the hygiene and quality of milk.
- Facilitating for wholesale and retail shops along the road and around the Bavi national eco-park.

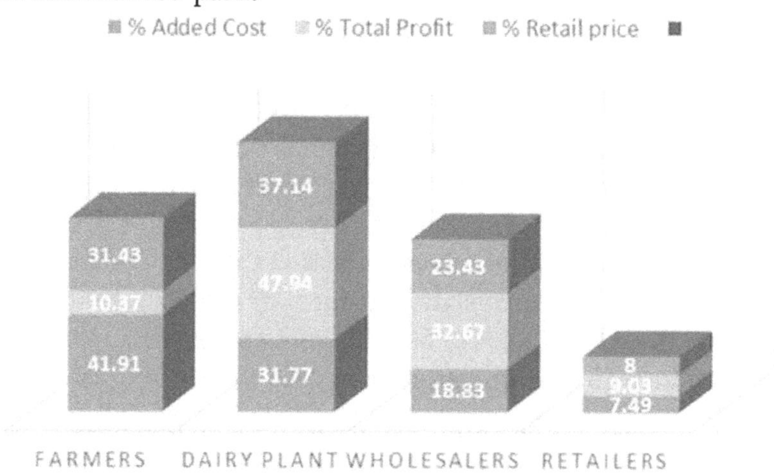

Figure 5. The comparison among actors in added cost, total profit and retail price-1. Source: Calculated by data collected, 2016.

Besides, there are the inevitable weak points at which express the lack of authority's control. Bavi milk brand so far was given right to only two companies. However, there are a lot of companies taking advantage of the lax management, they set up the companies with the same name to cheat consumers. They purchase milk from dairy farmers having under and poor standard milk; which affects the consumer health. Here, the list of the companies who were not allowed to use the brand of Bavi milk but still sell a huge amount of milk everyday (Table 9).

2.7. Concise Summary

Ultimately, income distribution in the chain is unequal. Farmers obtain insufficient advantages in comparison to the costs they incur. This specific chain is mostly driven by manufacturers, and the rising revenue directly correlates with the advantages they provide. As a result, the value-added in the chain is similarly skewed towards the dairy facility. This chapter highlighted the inadequacies in farmers' cost calculations. The overall cost of a dairy plant includes expenses like payroll and opportunity costs, which are not typically mentioned by dairy producers. Therefore, farmers experience greater disadvantages and less gains, despite the fact that they were supposed to earn higher incentives. Bavi's authorities were unable to efficiently control the milk production. The lax management has caused a proliferation of counterfeit items in today's market. The sale of substandard products at highway and tourist places in Bavi has a detrimental impact on Bavi's reputation, tourism industry, and consumer well-being.

Thus, firstly, it is necessary to make measures in order to obtain the balance among actors, especially dairy farmers.

- In the whole value chain, the dairy plants play the most important role. Farmers could not have enough power to negotiate with dairy companies about prices of raw milk. As a result, dairy companies could use many methods to buy raw milk of farmers at a low price. Thus, to increase the profits received of dairy farmers, it is indispensable to increase the role of farmers.

Table 9. Name of company illegally using Bavi milk brand.

Sr.	Name
1	Bavi Yogurt LTD company
2	Bavi milk investment JSC company
3	Vietnam Dairy—Bavi JSC
4	Fresh milk JSC company
5	Bavi fresh milk JSC company
6	Bavi milk and cake production JSC company
7	Bavi milk and cake JSC company
Source: Calculated by data collected, 2016.	

To ensure the criteria in strict quality of milk, farmers should choose the good breeding cows that have ability to avoid disease and get high productivity.

- The small-scale milk collection of dairy companies, in fact, is a costly method. Moreover, the gap in the education level of each factors spark difficulties for the implementation of contracts. Therefore, companies should combine small milk collection stations into the largest milk collection stations. The commune should establish some organizations to represent farmers to sign contracts with the companies.

- Building waste-treatment system: In Bavi, it is clear to the cattle waste in front of farmer's house (farmers aim to use it to grow grass). However, it not only pollutes the human living environment but also affect the cleanliness of cows and milk. Thus, it is necessary for government to build a waste-treatment system.

The above measures could not only solve the problem of distribution but also enhance the connection between actors. Management is also the problem that the thesis wants to give recommendations that improve the value chain of dairy products in Bavi.

- To be able to manage the value chain of Bavi milk, policymakers and milk entrepreneur should encourage factors to be aware of the value chain of dairy milk and the need and benefit when they participate deeply in the chain.
- Policymakersshouldcheckandreviewallenterprisesusing Bavi Milk brandfor developing their business. This activity could prevent other companies from using the Bavi milk brand for developing business illegally by taking advantage of the Bavi brand.
- Checking the quality of raw milk and milk products frequently. In serious problem, authorities should deprive the license registration of companies.
- Provide adequate information to dairy farmers about the companies operating illegally to avoid the milk supplying from farmers to those companies.
- Do not allow the distributors to sell the unclear brand product.

The implementation of these steps will enhance the Bavi milk brand, expanding its reach and ensuring that each participant in the value chain receives the appropriate advantages.

3. AT A GLANCE

Dairy production in Vietnam thrived in the early 1900s due to the region's ideal natural conditions. The Vietnam dairy sector has consistently developed and made major contributions to assuring food needs during challenging periods. Nevertheless, Vietnam's dairy sector has not been able to meet the country's milk consumption needs. In Vietnam, retail milk prices are exorbitant, but dairy producers sell milk at significantly lower prices. The cause originates from the regulation of dairy industries over the quantity and quality of milk. Furthermore, that control led to an unequal distribution of earnings and benefits among the participants in the dairy value chain. This study examines the distribution of benefits, costs, and value-added among the actors, as well as the challenges in the practical management of the dairy milk value chain, with a specific focus on Bavi as the case study.

REFERENCES

1. Khoi NV. Wicked problems: A value chain approach from Vietnam's dairy product. Springer Plus. 2013;2(1):1-6
2. Khoi NV. Global Value Chains of Transnational Corporations: The Practical Approach from China. Hanoi: VNU Publishing house; 2013
3. Porter ME. Competitive Advantage: Creating and Sustaining Superior Performance; 1985
4. Kapslin R, Morris M. A Handbook for Research Value Chain. Ottawa: IDRC; 2001
5. Khoi NV. The dairy industry in Vietnam: A value chain approach. International Journal of Managing Value and Supply Chains. 2014;5(3)
6. Khoi NV. Opportunities consolidation and mergers in the dairy sector in Vietnam from the perspective of the value chain. Journal of Economic Research (Economics Studies). 2014

07 STUDY OF SOYBEAN AGRIBUSINESS IN ARGENTINA FROM 1990 TO 2015: ANALYZING SOCIO-ECONOMIC, TERRITORIAL, ENVIRONMENTAL, AND POLITICAL CONSEQUENCES

1. INTRODUCTION: AN OVERVIEW

Geographic space can be understood through the analysis of the dynamics of the division of labor and the many uses of territory [1],. This involves examining the material and immaterial flows that demonstrate the spatial and temporal patterns of a certain cycle of capital's rotation and reproduction [2]. Agribusiness utilizes spatial circuits of production that are structured by agro-industrial value chains, encompassing various stages like agricultural input supply, rural activities[3],, industrialization, distribution of goods, transportation, financial operations, and commercial networks[4]. Conflicts and power dynamics arise in the process of appropriating surplus due to the interconnectedness of input-output relationships and the unequal power relations among the agents at each stage of the chain.

Each agribusiness cluster contains accumulation cores, leading capitalist agents to employ various tactics to organize the productive chain in alignment with their interests. Consequently, they seize the majority of the income generated by other parties and the resources provided by the State. This scenario suggests that agribusiness can be viewed as the rural manifestation of "neo-extractivism," which involves modernizing the exploitation of natural resources [5, 6] and the external appropriation of profits [7],that originated in Latin America during the colonial era and continued in subsequent periods.[8]. Soybean was originally considered exotic and rare in Argentina, but now it occupies 59% of the country's grain land and accounts for 93.2% of grain production. Soybean cultivation has increased significantly over the past four decades, expanding from 30,470 hectares in the agricultural year 1970/1971 to 206,025,042 hectares in 2015/2016. Soybean production surged by 451% from 10,671,100 tons in 1989/1990 to 58,800,498 tons in 2015/2016.[7, 9]. Consequently, Argentina is now the third largest producer and exporter of soybeans globally, and the leading supplier of soy oils and flours[10, 11].

Various factors contribute to explaining this change, including the compatibility of this crop with hybrid wheat production, the expansion of the European market for balanced animal feed, neoliberal structural reforms in the 1990s, the global livestock crisis, the continuous increase in external demand for vegetable proteins following the European Economic Community's approval of GM soybean imports and processing,[12] and the subsequent surge in international soybean prices. The agro fuels boom and the financial crisis in 2008 prompted investment money to shift from real estate markets to commodities markets, contributing to the consolidation of the soybean industry in Argentina.Although big farmers, agro-industrial enterprises, seed companies, and even the National State often invoke the alleged benefits of this pattern, during the last 15 years, a leafy scholarly literature has revealed the dark side of the soybean boom. Several researchers have presented general studies on the negative impacts of the soybean pattern [7, 13, 14, 15, 16, 17, 18, 19], whereas other authors exhaustively analyze individual features of this matter. For instance, Teubal emphasizes the vulnerable and dependant condition of soy monoculture and its export orientation [20], whereas Teubal and Rodríguez claim that soybean pattern is the major driver of the concentration and transnationalization of the Argentinean agro-industrial accumulation chain [21]. Hernández focuses on the cultural and empresarial fragmentation of the rural sector [22], while Domínguez and Sabatino state that soybean's boom has led to both the scarcity and the price increase of the national diet's traditional foods [23]. Other researchers have expressed concern about the risks of the transgenic soy agriculture at great scale, noting the loss of legal and food sovereignty stemming from the seeds' privatization [24, 25, 26, 27]. Furthermore, several papers have studied the environmental implications of soybean fever, specially soil erosion, virtual water exports, and deforestation [28, 29, 30, 31, 32]. Rodríguez Striebeck examines the impact of the soybean boom on land disputes. Other studies have examined the connection between soybean fever, fumigations with agrochemicals, and the significant increase in oncological disorders and cases of abnormalities in newborns in rural areas in recent years [34, 35, 36, 37, 38].

GM soybean is a prominent representation of agribusiness in Argentina and is considered one of the most contentious and problematic land uses in the country. The major objective of this chapter is to analyze the socio-economic, territorial, environmental, and political ramifications resulting from the expansion of the soybean pattern during the past 25 years.

Analyzing the socio-economic issues of the soybean agribusiness involves studying the technique and productive restructuration of the rural sector, new leasing practices, and the vertical integration of the agro-industrial chain. This analysis includes examining changes in the area sown with other crops and occupied by livestock, the reduction of rural units, the surface leased to sowing pools and investment funds, the origin of capital, and the involvement of major soybean firms in various stages of the accumulation chain. Secondly, the territorial perspective focuses on the new enclosures resulting from the concentration of land and the displacement of peasants and aborigines. This is illustrated by examining the land acquired by landowners, conflicts over land in soybean-producing regions, and the methods of dispossession employed by farmers, agricultural investment funds, and agro-industries. The text examines the environmental and health impacts of the expansion of the soybean frontier, including deforestation of native forests, ecological degradation such as virtual water exports, soil erosion, and greenhouse gas emissions, as well as the use of agrochemicals in soybean cultivation and health issues related to fumigations. The impact of the global seed industry on farmers' control and the loss of legal and food sovereignty is shown by analyzing the leading companies in the sector, outlining their strategies for growth, and detailing conflicts related to the purchase and utilization of transgenic soybean seeds.

2. THE SOCIO-ECONOMIC IMPACTS OF THE SOYBEAN BOOM INCLUDE AGRICULTURAL RESTRUCTURING, DECLINING RURAL UNITS, NEW LEASING METHODS, AND VERTICAL INTEGRATION OF PROPERTY OWNERS.

Soybean agribusiness mostly focuses on exporting and accumulating products in a chain-like process. Over 90% of the production is sold to nations such as China, India, the European Union, Pakistan, Bangladesh, and Japan [10, 11]. Since soybean value chain concentrates 40% of agricultural exports [12] and is the major exporter sector (27.7%) of the Argentinean economy [39], such productive circuit generates bulky profits (20,000 million of dollars per year). A substantial part (30%) of these profits is appropriated by the National State by collecting taxes on exports. Nevertheless, the soybean value chain's contribution to labor market is very small. Actually, the labor force employed in the soybean cluster reaches 193,894 workers, i.e., 10.3% of agro-industrial employment and 3.1% of whole employment. Such figures imply a labor intensity (seven jobs per million of dollars of value), three times smaller than the national average [40, 41]. In addition, soybean boom has led to a dramatic restructuration of the rural sector in our country.

As shown by Figure 1, agricultural stage of the soybean circuit has acquired a noticeable geographic dispersion, not only strengthening in the Pampas, i.e., Buenos Aires, Córdoba, Santa Fe, Entre Ríos, and La Pampa, but spreading much of other Argentinean provinces (e.g., San Luis, Jujuy, Salta, Catamarca, Santiago del Estero, Tucumán, Chaco, Formosa, Corrientes, and Misiones). In fact, all these provinces increased its area planted with soybean among 33.3 and 7.920% during the period 1990–2015, except for Misiones [9]. According to the last Agricultural National Census performed in 2008, soybean has absorbed among the third and more than half of the agricultural area of the provinces of Buenos Aires, Santiago del Estero, Chaco, Tucumán, Córdoba, Entre Ríos, and Santa Fe [42]. Significantly, soybean means 66% of the land conquered by the agricultural frontier's advance in Chaco, Tucumán, Salta, and Santiago del Estero [11]. Although both the Pampas' periphery (e.g., Entre Ríos, and La Pampa) and the provinces located outside this region (e.g., Chaco, Santiago del Estero, etc.) have increased substantially its relative weight on the planted area, the Pampas' core still concentrates 85% of soybean's production units [9, 11]. Likewise, the chain's agricultural stage shows high levels of business concentration. In fact, 6% of agricultural units mean 54% of soybean grain's production [43].

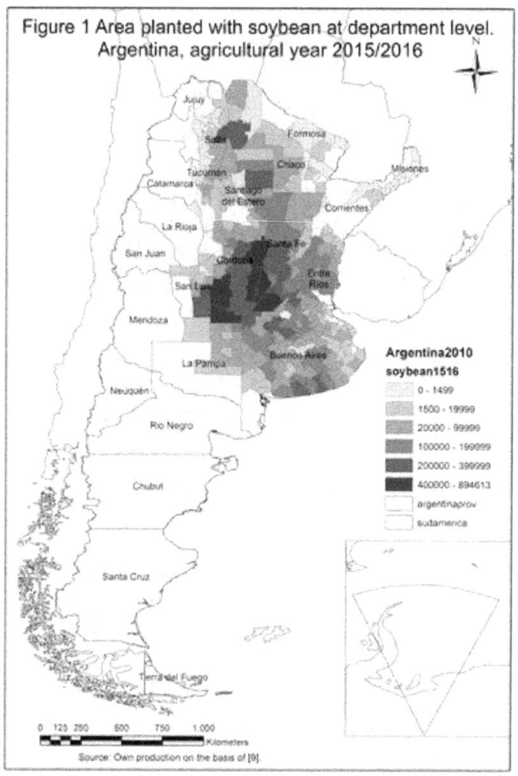

Figure 1.Area planted with soybean at department level. Argentina, agricultural year 2015/2016.

According to Silveira, nowadays modern uses of territory seek to erase the traces of the past in order to the space be rewritten by the new rationalities of the present [44]. The aforementioned soybean expansion is, by way, a clear example of this logic, though it has led to the disappearance of both the ancient practices of rotation among stockbreeding and agriculture, and the traditional regimes of alternation between diverse crops. For instance, more than 10 millions of heads of cattle were suppressed during the 1990s, while the livestock decreased 6% between 1988 and 2002 and the amount of milking yards was halved [23]. In the Pampas, i.e., a strongly milk's production-specialized region, some milk companies closed most of their production units to plant soybean. Furthermore, scarcity of lands for stockbreeding has implied decreasing of the livestock area, as well as the cows' concentration in the so-called feed lots.

Other traditional productions were sacrificed in order to free lands to the soybean frontier's advance. For example, the oats area decreased by 37% between 1989/1990 and 2015/2016, as well as sunflower and barley area reduced by 48.7 and 50.5%, respectively. Moreover, the falls suffered by the canary grass (66.8%), the millet (87.7%), and the flax areas (97.1%) were even more substantial [9]. Generally speaking, although other crops have increased its area (e.g., sorghum, corn, wheat, bread wheat, horticultural products, peanut, etc.), such global variations hide a dramatic reduction during the early years of soybean boom. Outside the Pampas, the previous structural crisis of some crops was functional to the monoculture's advance too. For instance, Tucumán lost the quarter of sugar cane area between 1990 and 2001, whereas the Chaco's cotton area reduced by 52.3% during the period 1990–2015 [9].

Owing to the growing weight of science and technology in rural activities, the new soybean production's requirements have involved the introduction of specialized technical systems (e.g., direct sowing, modern rural machinery, global-action herbicides, fertilizing with nitrogenous, prilled urea and phosphorus, etc.). Both the modernization of agricultural practices and the incorporation of the last technology led to the massive indebtedness of the rural producers. Due to the high interest rates, the original debts' values increased 12 times in few years. As a result, in the beginning of the 2000s, the rural sector owed 7000 million dollars to banks, as well as 3000 million dollars to inputs suppliers. Such situation implied the bargain sale of 10 million hectares and the mortgage of 14 million hectares [7, 12]. Since they were unable to resist the productive and financial costs of the new soybean's pattern, several small and medium-size producers disappeared, the same ones who could renew their equipment and living with dignity before the soybean boom [45].

According to the Agricultural National Census, almost third (32.9%) of the Argentinean rural production units disappeared during the period 1988–2008 [42, 46]. A strong, noticeable empirical correlation between such farms reduction and the soybean specialization may be identified. Significantly, none of the 10 major soybean provinces lost less than the quarter of rural production units during the analyzed period, except for La Pampa and Salta. In addition, these ten provinces represent 68.8% of the missing farms of the country (95,418 over a total of 138,639). Such reduction was equal to or higher than the national average in almost half of the cases, as demonstrated by Entre Ríos (-32.8%), Córdoba (-33.7%), San Luis (-38.6%), Tucumán (-53.7%), and Buenos Aires (-54.1%). Finally, the four major soybean provinces, i.e., Buenos Aires, Córdoba, Santa Fe, and Entre Ríos, have meant 53.1% of missing units of rural production.

Most of those small- and medium-size farmers who kept their lands finally leased it to extra-agricultural origin companies, such as sowing pools and investment funds. These companies include a great variety of firms, such as agronomies, big storing enterprises, input suppliers, banks, insurance companies, managers of retirement and pension funds, isolated investors, rural machinery contractors, agro-industrial companies, agricultural producers, writing desks, and both national and foreign holdings [23]. Although the sowing pools' origin is very diverse, such leasing practices are basically carried out by the financial capital, which resorts to various mechanisms (e.g., trusts, foreign funds capitalizations, stock-exchange organizational forms, local alliances, informal agreements, etc.) to appropriate the rural lands in order to control the accumulation of the chain's primary stage and entering and exiting the commodities market quickly [23, 47]. Currently, 70% of agricultural area of the country has been leased to sowing pools and investment funds [17].

Although all these capitals resort to different mechanisms (e.g., leasing-network firms, patrimonial ownership, purchase of lands by using external capitalization funds, etc.), they develop a common strategy: diversifying agro-weather, economic, political, and legal risks in order to combine high levels of liquidity, rapid returns on initial investment, and big financial and agricultural rents. Thus, such mega-companies obtain lower prices on the inputs supply, impose conditions to the farmers, and influence the prices of the land [47].

As a result, a lot of small- and medium-size farmers have been relegated from the productive circuit, keeping the nominal property of their fields but losing all control over the agricultural practices developed there. Provided the economic rationality of sowing pools is obtaining the highest profit in the short terms, farmers' lands often are super-exploited. Once farmers had regained effective possession of their lands, they face several difficulties in order to re-enter the productive circuit due to the loss of fertility, the destruction of soils, and the high economic costs of its reparation. Consequently, farmers use to sell their lands, thus increasing the concentration of rural property in few hands.

Typical agro-industrial backward integration strategies are developed too. Agro-industrial stage of soybean circuit includes the production of flours and oils, as well as the storage and shipment of these products. As the agricultural link, the agro-industrial stage is highly concentrated, both in economic and territorial terms. In fact, 80% of crushing plants are mainly located in the south of Santa Fe, i.e., the biggest port and agro-industrial cluster in the world. As the great crushing capacity of the Argentinean soybean cluster is very similar to the USA and Brazil ones, oil industries feed not only from the national production but the grains imports from Paraguay. Soybean agro-industry has an oligopsony structure, which includes the major grain traders from USA, France, Switzerland, Netherlands, and China (e.g., Cargill, Louis Dreyfus, Glencore, Bunge, Noble, Nidera, and Cofco). However, some big national firms have a significant participation in soybean agribusiness too, such as Oleaginosa Moreno, Aceitera General Deheza, MolinosCañuelas, Vicentín, and Molinos Río de la Plata. Remarkably, only thirteen companies produced 89% of both soybean flours and oils in 2015, while just eight firms meant 76.1% of oils and 82.4% of flours [48].

Moreover, the corporations aforementioned control two other strategic links of soybean circuit: the biodiesel production and the exports. Owing to both the world agro fuels boom and the growth in European and US markets, the Argentinean biodiesel production increased 435.7% between 2007 and 2010, rising from 560,000 to 3,000,000 tons. The domestic biodiesel market is mainly controlled by Louis Dreyfus, Eurnekián, Bunge, Aceitera General Deheza, Vicentín, Glencore, and Molinos Río de la Plata [49]. Regarding the chain's final stage, soybean agribusiness currently represents both 70% of the Santa Fe's foreign trade and between the third and the half of Córdoba's exports. Furthermore, soybean means around the tenth, the fifth and the quarter of the exports of Buenos Aires, Entre Ríos, and La Pampa, respectively. Finally, this value chain represents 25% of the foreign trade of Santiago del Estero and Chaco [50].

In addition, it is worthwhile noting that some oil industries and grain traders (e.g., Louis Dreyfus, Bunge, Nidera, Aceitera General Deheza, and Molinos Río de la Plata) have purchase and/or leased large tracts of rural lands for soybean sowing in order to ensure a stable supply of material raw [51]. Furthermore, these agro-industries have created their own storing territorial networks and so have reinforced their control over the circuit's primary stage. The aforementioned firms and others companies (e.g., Vicentín, Cofco, Toepfer, Cargill, ADM, etc.) have carried out forward integration too, specifically toward the logistic and transportation stages. This has been done by using three mechanisms: the purchase of big fleets of trucks, the control of the share capital of some railroads (e.g., The New Central Argentinean Railroad, which belongs to Aceitera General Deheza), and the appropriation of most of the port infrastructure of the provinces of Santa Fe, Buenos Aires, Entre Ríos, and Chaco.

3. TERRITORIAL PERSPECTIVE: SOYBEAN FEVER, RURAL PROPERTY'S CONCENTRATION, AND THE NEW ENCLOSURES

Soybean area's growth and the land's concentration are two phenomenon closely related. In fact, 936 landowners hoarded 35,515,000 hectares in the beginning of the 2000s, whereas 137,021 farmers only gathered 2,288,000 hectares [14]. Moreover, just nine holdings control 32% of rural lands of the province of Buenos Aires. This is because of two factors. On one hand, traditional rural families have joined foreign capitals by complex mechanisms (e.g., trusts, anonymous partnerships, dummies, farms subdivisions, fake sales, etc.) in order to shirk the provincial Treasury [52]. On the other hand, some mega-companies have emerged too, as shown by the cases of both national (e.g., Los Grobo, El Tejar, MSU, etc.) and foreign firms–for instance, Cresud and Adecoagro, belonging to the Hungarian-American tycoon George Soros. According to various estimations, such mega-companies control an area of 1,500,000 hectares [7, 12].

It is worthwhile noting that both land's concentration and new leasing practices have strongly relied on the soybean's expansion outside the Pampas' core. Operating as the driver of a substantial worsening of rural violence, such phenomenon is clearly associated to the so-called new enclosures, i.e., the silent but relentless cornering process suffered by peasants and aborigines during the last 2 decades [53]. Since the soybean fever encouraged big farmers, landowners, sowing pools, investment funds, and oil industries to expand toward the Pampas' edge and the Argentinean North, new territorial enclosures have risen in such regions. Current legislation both recognizes Aboriginal land rights and ensures the land ownership to those peasants who can prove they have occupied and exploited it during at least 20 years. Nevertheless, the legal precariousness of the land property in various provinces often implies that subaltern groups to be easily overwhelmed by the hegemonic agents of soybean agribusiness.

According to official data, 857 conflicts on the control of the land were reported in Argentina in 2011, involving 63,843 rural families and an area of 9,293,234 hectares. It is worthwhile noting that 48.5% of these situations were denounced in soybean provinces (e.g., Buenos Aires, Chaco, Córdoba, Entre Ríos, Formosa, La Pampa, Salta, San Luis, Santiago del Estero, and Tucumán) [54]. In addition, unofficial sources claim that land struggles have been increased in the Argentinean North and the Pampas' edge during the last years. In fact, 224 conflicts were reported en 2013, which involved 2,791,302 hectares and 127,886 people. Significantly, 80% of such conflicts started since 2000 [55], i.e., since the soybean's expansion outside the Pampas sped up.

A leafy scholarly literature has reported a wide range of coercion mechanisms used in order to loot and misappropriate the peasants' and aborigines' lands. On one hand, some manipulation and deception strategies allow farmers and indigenous to be evicted without major difficulties (e.g., signing blank documents that actually are lending or eviction agreements) [56, 57]. On the other hand, political and legal power operate as dispossession guarantors by both ordering evictions for indigenous and peasants and accepting the fake company-owned documents as legal. Such support often involves the entire repressive apparatus of the State, from judges and lawyers to congressmen and public forces of security, who grab the rural families' lands to put them under the control of Buenos Aires, Córdoba, Santa Fe, and Tucumán's soybean companies [52, 57].

Finally, dispossession of the land implies nonlegal actions too, even paramilitarization. Physical wounds and even death are the result of territorial struggles unchained by the agribusiness logic [53]. Several strategies are commonly used on this regard, such as de facto occupations, even when peasants have got legal documentation that legitimized their land rights, land enclosures and deforestation, and blocking access to roads, grazing areas, and drinking water sources. Furthermore, the so-called white guards, i.e., civilian paid by the soybean producers, intentionally pollute water wells, kill livestock and farm animals and pets, threat farmers, fire ranches, and even murder peasants and aborigines [45]. One of the most affected provinces is Santiago del Estero, where 122 conflicts were reported in 2013, over a total of 224 [55]. The misappropriation of almost 4 million hectares in this province has implied evictions, violations of human rights, and even deaths. Often real states agencies sell lands with its inhabitants inside, then vacating and guarding them by using provincial police's special brigades. Such complicity chain includes entrepreneurs, lawyers, notaries, judges, and congressmen too [52, 55]. Similar practices have been reported in Salta [58].

4. ENVIRONMENTAL AND SOCIO-ECOLOGICAL ISSUES: DEFORESTATION, DEGRADATION, AND THE HEALTH QUESTION

Silent enclosures and undisguised strategies of violence have been complemented by deforestation. Almost 5 million hectares of native forests were cut down in just 16 years (1998–2014) [59], a figure that means 4.3% of world deforestation, according to the International Panel on Climate Change [57]. In fact, the national annual rates of deforestation fluctuate from 0.49 to 1.31%, i.e., 9.4 times bigger than the world annual averages (0.22 to 0.14%) estimated during the period 1990–2010 [59, 60, 61, 62]. Although this serious question led to deforestation was restricted and even banned by the State, illegal logging continues because of the unceasing expansion of the agricultural frontier [57].

Soybean's advance has been the major driver of forest area reduction between 2002 and 2011 [59], as well as an indirect deforestation driver too, though it has expelled traditional productions from its historical cores (e.g., cotton, sugar cane, stockbreeding, etc.). Remarkably, 95.8% of cleared area (4,754,747 hectares) involved provinces that belong to both the soybean core's area and the peripheral regions recently conquered by the soybean's expansion. As shown by Figure 2, deforestation map mainly involves soybean provinces, such as Santiago del Estero, Salta, Chaco, Córdoba, Entre Ríos, and Formosa, which concentrate 85.2% of the logged area in the whole country. Moreover, such provinces have substantially exceeded both the national and world averages. For instance, the deforestation rate of the north of Córdoba during the period 1998–2006 was 13 times bigger than the world rate, whereas the Santiago del Estero's logging speed during the period 2006–2014 was 7–19 times bigger than the international average. Finally, Salta is currently known as the "capital of logging." This is because the government uses to modify the regionalization established by the Forests Law in order to satisfy the soybean firms' requirements. In addition, foremen, corporations, and organisms strongly pressure aborigine communities by demanding them to sign deforestation permits or to accept the logging of conservation areas in exchange of water, food, and ambulances [57].

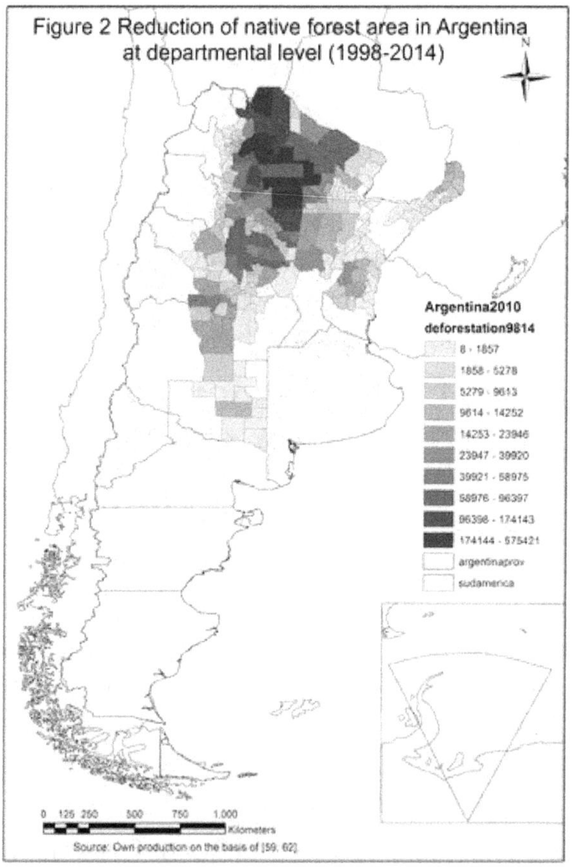

Figure 2.Reduction of native forest area in Argentina at departmental level (1998–2014).

Soybean's boom is also one of the major responsibility for the growth of greenhouse gases' emissions, the loss of biodiversity, and the increasing frequency of floods and rockslides in our country. Likewise, soybean is the main crop involved in desertification process, due to the for free-exports of virtual water and essential edaphic resources (e.g., nitrogenous, phosphor, potassium, calcium, magnesium, sulfur, iron, manganese, boron, zinc, molybdenum, chlorine, copper, etc.) [29, 30]. Environmental degradation (specially, deforestation) completes the expulsion of Aborigines by destroying their ethnical, cultural, and socio-economic subsistence matrix. Indigenous population of the Argentinean North was estimated in 900,000 people in the beginning of the 2000s, the half of whom have been condemned by deforestation to begin big and medium-size cities of the provinces of Santa Fe, Chaco, Formosa, Salta, Misiones, and Buenos Aires [14]. Thickening the extreme misery and poverty's urban belts, such masses of dispossessed often suffer starvation and restricted access to drinking water, as well as high levels of child mortality due to the migration of zoonotic illnesses (e.g., hanta virus, dengue, leishmaniasis, etc.) from the forest to the cities [58].

Another serious implication of soybean agribusiness is the health issue caused by the massive fumigations with agrochemicals. National agrochemical consumption has exponentially grown since the introduction of transgenic soybean and its technological package, climbing from 39 liters in 1991 to 335 million of liters in 2012. While in other countries the use of agrochemicals does not use to exceed 3 liters per hectare, the national average is 12 liters per hectare, reaching peaks of 20 liters in Santiago del Estero and Chaco. Such values would imply that each Argentinean inhabitant is exposed to a dose of 8 liters of agrochemicals, whereas this figure would oscillate from 30 to 60 liters per capita in the soybean regions [63]. Soybean agrochemicals' package includes glyphosate of ammonium and other global-action herbicides, insecticides, and fungicides, such as cypermethrin, chlorpyrifos, paraquat, glufosinate of ammonium, bromoxynil, malathion, 2,4-D, endosulfan, hexachlorobenzene, heptachlor, and atrazina. It is worthwhile noting that the last five agrochemicals mentioned above have been banned or restricted in several countries, whereas glyphosate of ammonium has been classified as "probably carcinogenic" by the World Health Organization in 2015.

According to the Atlas of Childhood's Environmental Risk in Argentina, the most affected departments by the use of pesticides largely matches regions that have been intensely conquered by the soybean monoculture, i.e., center and southeast of Córdoba, south of Santa Fe and Chaco, north of Buenos Aires, much of Santiago del Estero, Tucumán, and Entre Ríos, etc. (Figure 3). Soybean, in fact, is the main crop involved in the Pollution by Pesticides Index of the mentioned Atlas [64].

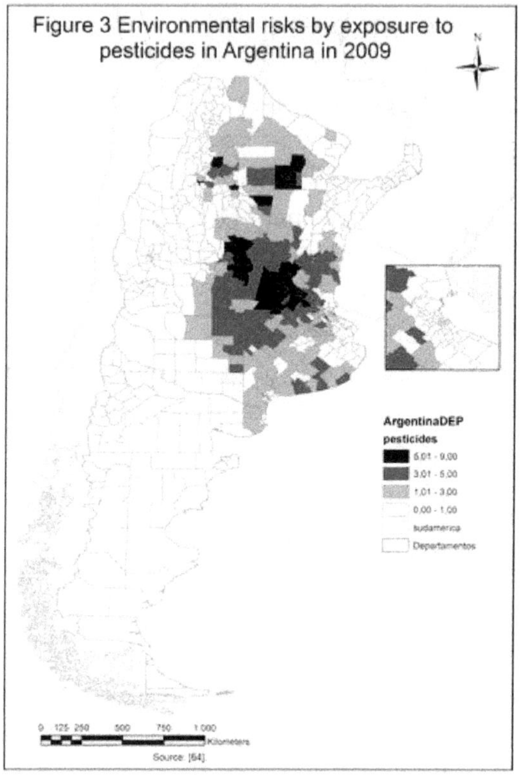

Figure 3. Environmental risks by exposure to pesticides in Argentina in 2009.

Soybean regions' inhabitants have reported a wide range of serious socio-economic, environmental, and health damages due to the aerial and land spraying with agrochemicals and the pollution produced by grain and pesticides' storing plants. Some of these damages have included the hives and horticultural cultivations' destruction, the mortality or/and degenerative disease of poultry and cattle, the pollution of air, soil, and surface and groundwater, and certain health issues (e.g., chronic respiratory diseases, diarrhea, dizziness, cephalalgia, nauseas, vision problems, dermal rashes, hormonal disruptions, and thyroid disorders). Even more serious pathologies have been attributed to the soybean's pattern too, such as hemolytic anemia, rheumatoid arthritis, lupus, purpura, central nervous system damage, Hodgkin's disease, multiple sclerosis, cerebral ischemia, toxic liver diseases, and neurological disorders. Furthermore, other health issues have been reported in the Pampas and the Argentinean North, such as male sterility, births with congenital malformations, mutagenesis, and spontaneous abortions. Finally, even some kinds of cancer (e.g., leukemia, and urogenital, pancreatic, lung, breast, prostate, stomach, and liver cancers) have spread among the rural workers and the people living near crops and storing plants. Importantly, in soybean regions both the malformations and cancer's incidence is 2 to 10 times bigger than the national and urban averages [57, 63, 65].

5. POLITICAL IMPLICATIONS: SEED AND AGROCHEMICAL COMPANIES, FARMERS' AUTONOMY, AND LEGAL AND FOOD SOVEREIGNTY

Despite the networks paradigm would seem to hide deep inequalities among the different agribusiness' agents and links, transnational corporations actually controls the whole global value chains by defining both the general parameters of the agro-industrial system and commanding the technological innovation's diffusion. During the last 20 years, the so-called transgenic revolution encouraged seed companies to develop a complex mergers and acquisitions process, which involved agrochemical and even pharmaceutical firms too. Owing to such oligopoly structure, a handful of corporations (Monsanto, DuPont, Dow Agrochemical, Bayer, Basf, Aventis, and Syngenta) control 100% of transgenic seeds market and 97.8% of agrochemical market at world scale [14, 26].

Such concentration of power allows companies to appropriate the so-called life's rent [45] and to deploy several subordination and dependence mechanisms by applying owner rights on seeds (e.g., patents, licenses, etc.) and creating new biotech packages (e.g., when a novel seed's genetic feature and an agrochemical input are combined, patented, and sold by the same company). As a consequence, farmers are prevented from easily reproducing the new seeds and forced to acquire certain pesticides and fertilizers, thus losing all the control over the first chain's link and becoming a captive market (simple germplasm's leaseholders) for the companies [14, 26].

Such is the case of the Argentinean soybean agribusiness, which are based on two inputs patented and marketed by Monsanto at world scale: the RR glyphosate-resistant soy and the glyphosate of ammonium. However, since the government approval of this crop variety only occurred in 1996, the initial, fast expansion of the new seed during the earlier years of soybean boom was due to genetic smuggling and clandestine plantations, two phenomenon actively encouraged by Monsanto [66]. As a consequence, nowadays 99.1% of the soybean planted in Argentina is transgenic [7].

From the mid-1990s onwards Monsanto has became the main agent of the whole soybean accumulation chain by using different mechanisms. On one hand, the US corporation controls both seed and agrochemical markets under almost monopolistic conditions. For instance, Nidera, the leading company of the Argentinean Seeders Association (ASA), operates as a Monsanto's license firm by marketing transgenic soybean seeds belonging to this US company [14]. Such situation has been reinforced by the Magnum Network, a system created by Monsanto in 1998 that imposes an exclusivity contract on agronomies. Owing to that exclusivity contract, agronomies cannot sell seeds of competitor firms and must accept both to adopt direct billing conditions and to provide detailed customer information by using a management system directly connected to Monsanto's sales and marketing management [12].

Likewise, the Argentinean agrochemical market brings a return of 2.5 billion dollars per year to the companies, 64% of which corresponds to glyphosate of ammonium [63]. Although such pesticide is sold not only by Monsanto, but other companies too (e.g., Syngenta, Basf, Bayer, DuPont, Dow AgroSciences, Nidera, Atanor, La Tijereta, etc.) [67], the US corporation is the one who appropriates the most part of the bulky profits generated by the sale of glyphosate. Finally, Monsanto is the Cargill's share capital owner, thus controlling a large part of the agro-industrial stage [24].

In the beginning of the Argentinean transgenic soy boom Monsanto did not strive to directly benefit from the sale of the RR seed, but focused its attention in the marketing of glyphosate [12]. As a consequence, farmers started to develop practices legally recognized by both the national legislation and the FAO's Phytogenetic Agreement, such as using the seed free of charge, sharing it with other producers, multiplying it, and even selling and/or saving it for later re-sowing. Nonetheless, Nidera and other license companies obtained in 1999, the government approval to impose an "extended royalties" system. According to that system, when farmers acquire the RR soybean seed automatically sign a contract that not only impedes them saving and/or sharing their harvest, but forces them to pay 2 dollars (plus taxes) for each 50 kilograms-bag of certified seeds they keep for planting in the following agricultural year [15].

Since such contract meant a noticeable transgression to the Argentinean Seeds Law, the extended royalties system was very resisted by the soybean producers, as well as avoided by various mechanisms. As a result, in the beginning of the 2000s, the use of certified seeds represented less than 18% of the soybean planted [66]. However, Monsanto only start the relevant formal claims when the spread of its transgenic soy in the country was absolute, its Magnum Network was already consolidated [12], and the allegedly "clandestine" practices of Argentinean farmers had favored it with the dissemination of the RR soy in Paraguay, Brazil, and Bolivia, i.e., countries where transgenic crops were still banned [66]. By using the pretext that farmers had violated the legal security of the seed industry and injured their intellectual property rights, the US company begun to deploy different mechanisms of pressure and lobbying on the Argentinean government [15].

Monsanto's first strategy was activating the royalties payment by the soybean marketed in Argentina. Due to the failure of this tactic, Monsanto asked for national government to sanction a "global royalties' law." Such law should create a Technological Compensation Fund oriented for collecting an aliquot that fluctuated

0.35–0.95% of selling price of soybean and lately distributing such resources among the seed companies [15]. Since the bitter resistance of large farmers determined the unfeasibility of the legislative project, Monsanto implemented other measures in 2005. By demanding a much higher fee (15% per ton) than the traditional royalty, the US company initiated litigations against Argentina in international courts and temporarily abandoned the commercialization of soybeans in the country [66]. Moreover, Monsanto fomented export blockades, achieved that European courts stopped and confiscated ships loaded with Argentinean soybean in the Old Continent's ports, and soughed the support of congressmen, ambassadors, and politicians of the US government [57]. Although Monsanto possesses the monopoly on RR soy at world scale, such right is not valid in Argentina, since the company never patented the transgenic event there [66].

Owing to the poor results obtained by these strategies, Monsanto introduced another mechanism of coercion: controlling the later diffusion of technological change. In 2012, the US corporation patented the RR2 Intacta Pro soybean, which is not only tolerant to glyphosate (like the conventional RR variety), but also has the Bt gene–it is resistant to some insect pests. Nevertheless, Monsanto conditioned the commercialization of the new transgenic event to the production of legal frameworks (both public and private) functional to its interests. In fact, the company stipulated that its new soybean could only be acquired by those farmers who previously signed a license agreement and expressly agreed to pay "extended royalties" at the time of purchase [57].

As such system was much resisted by the soybean producers, in 2016 the National State finally decided to regulate the seed market and operate as a guarantor of the Monsanto's intellectual property rights. On the one hand, the government confirmed the validity of the contracts previously signed between the company and the farmers. On the other hand, the State determined that the control of the origin (legal or illegal) of the seed should be carried out by the National Seed Institute (INASE). If the RR2 soybean's origin is illegal, INASE will apply sanctions and allow the seed companies to make payment claims to producers, around 12 dollars per ton [68]. Finally, this situation would be reinforced by the imminent sanction of a new law of seeds, which will establish that only small farmers will be able to keep seeds for free for later use, whereas larger farmers will be forced to tax companies.

6. CONCISE SUMMARY

The Argentinean soybean agribusiness offers numerous empirical examples that illustrate the conflicts and mechanisms of domination and subordination within and outside agricultural production circuits, as revealed by accumulation chains showing the spatial aspects of capital rotation and reproduction. The soybean boom has significantly impacted the national rural sector, leading to a historical turning point. This has resulted in a paradox where the expansion of soybean cultivation, covering over half of the grain area and breaking harvest records annually, is causing a decrease in the number of farms.

Three accumulation centers can be recognized in the soybean industry. The first refers to the elite class of major agricultural production units in the first period. However, the distinctions between this sector and other parts of the agribusiness chain are becoming less clear due to a new phenomenon where non-agricultural agents and field tenants, rather than owners, are gaining significant influence in primary production. Soybean sowing pools are conglomerates where many capitalist forces come together, merging new leasing procedures with traditional land acquisition strategies.

The primary focus of the second soybean chain is the oil and flour industry, which is marked by significant levels of consolidation and foreign ownership. The agro-industry has implemented many strategies for backward and forward integration. Oil companies have bought and/or rented extensive areas of land to guarantee a consistent supply of raw materials at a low price, so strengthening their dominance over the initial phase of the production process. Furthermore, these corporations have diversified into other phases of the supply chain, including biodiesel production, and have control over both transportation and international trading routes. Due to the accumulation methods of both sides, rural property is increasingly concentrated in a few hands, leading to the erosion of autonomy for small and medium-sized farmers. Significantly, this occurrence is also associated with the creation of new territorial boundaries in the outskirts of the Pampas and the northern region of Argentina. Agro-industries, landowners, and sowing pools use deceit, manipulation, usurpation, eviction, blockade, repression, intimidation, and assassination to seize land and force indigenous people and farmers to relocate to impoverished urban areas. Soybean production causes environmental destruction such as deforestation, loss of biodiversity, soil and water depletion, and desertification.

It also leads to serious health issues for people living near soybean fields, plants, and storage facilities due to pesticide spraying and agro-industrial pollution. The State profits significantly by imposing export taxes on soybean income, which allows it to observe and potentially support processes of oppression and destruction. The third accumulation core refers to the global seed and agrochemical sector. This industry controls the entire soybean agribusiness through various mechanisms such as networks of license firms, diffusion of technological change, ownership of a major grain trading company, intellectual property royalties, and political power. The accumulation strategies have increased farmers' reliance and diminished their legal and food sovereignty. This has resulted in a hidden hierarchy of subordination and exploitation that impacts even the major agricultural producers, leading to power struggles over profit distribution. Political control and economic surplus in the soybean chain concentrate in the phases (seed and agrochemical companies, oil and flour sectors) that contribute the least to value creation, with percentages of 8 and 10%, respectively, compared to 73% in primary production [41]. In conclusion, the consequences discussed, such as the minimal impact on employment generation, indicate that the soybean value chain should be viewed as an inefficient use of land for the majority of the country's population.

7. AT A GLANCE

Nowadays, soybean value chain is both the major expression of agribusiness and one of the most troublesome uses of territory of Argentina. This chapter is aimed to analyzing the worrying socio-economic, territorial, environmental, and political implications unchained by the expansion of the soybean's pattern during the last 25 years. On the basis of scholarly literature and both official and unofficial sources of data, we have studied the restructuration of the rural sector, the concentration of both the rural property and the agro-industrial chain, the new territorial enclosures, the socio-ecological and health consequences of the soybean's advance, and the influence of the transnational seed industry on the farmers' subordination. Our results show a substantial reduction of both the amount of rural units and the traditional production areas, the emergence of new leasing practices, the accumulation chain's vertical integration, the growth of the land's concentration, the expulsion of aborigines and peasants, the increase of deforestation and environmental degradation, the loss of legal and food sovereignty, and the serious impacts on the population's health due to the massive fumigations with agrochemicals. The chapter's findings suggest that soybean agribusiness should be considered as an irrational use of territory for most of the national society.

REFERENCES

1. Santos M, Silveira ML. O Brasil. Território e sociedade no início do século XXI. Rio de Janeiro-São Paulo: Record; 2001. 471 p
2. Lobato Corrêa R. Interaçôesespaciais. In: De Castro IE, Da Costa Gomes PC, Lobato Corrêa R, editors. Exploraçôesgeográficas. Percursos no fim do século. São Paulo: Bertrand; 1997. pp. 279-318
3. Austin J. Agribusiness in Latin America: A Case Book on Agribusiness Management. New York: Praeger; 1974. 322 p
4. Moraes ACR. Los circuitosespaciales de producción y loscírculos de cooperaciónenelespacio. In: Liberali AM, Yanes L, editors. Aportes para elestudio del espaciosocioeconómico (III). Buenos Aires: El Coloquio; 1986. pp. 151-177

5. Cariola C, Lacabana M. Circuitos de acumulación: unaperspectiva de análisis integral para la planificación regional. Cuadernos del CENDES 5 Planificaciónen Venezuela. 1986;5(2):65-110
6. Rofman A. Las economíasregionales a fines del siglo XX. Los circuitos del petróleo, del carbón y del azúcar. Buenos Aires: Ariel; 1999. 240 p
7. Palmisano T. Los señores de la tierra. Transformacioneseconómicas, productivas y discursivasenelmundo del agrobonaerense [thesis]. Buenos Aires: UBA; 2014
8. Gudynas E. Diez tesisurgentessobreel nuevo extractivismo. Contextos y demandas bajo elprogresismosudamericano actual. In: Schuldt J, Acosta A, Barandiarán A, Bebbington, A, Folchi M, CEDLA, Alayza A, Gudynas E, editors. Extractivismo, política y sociedad. Quito: CAAP-CLAES; 2009. pp. 187-225
9. MINAGRI. Estimacionesagrícolas. Series históricas. Buenos Aires: Ministerio de Agroindustria; 2016. Available from: https://datos.magyp.gob.ar/reportes.php?reporte=Estimaciones
10. Pierri J. El boom de la soja. Un retorno al pasado. RealidadEconómica. 2006;219:53-65
11. Giancola SI, Salvador ML, Covacevich M, Iturrioz G. Análisis de la cadena de soja en la Argentina. EstudiosSocioeconómicos de losSistemasAgroalimentarios y Agroindustriales N° 3. Buenos Aires: INTA; 2009. 119 p
12. Gras C, Hernández V. Radiografía del nuevo campo argentino. Del terrateniente al empresario transnacional. Buenos Aires: SigloVeintiuno; 2016. 299 p
13. Giarracca N, Teubal M. El campo argentinoen la encrucijada: estrategias y resistenciassociales, ecosen la ciudad. Buenos Aires: Alianza Editorial; 2005. 514 p
14. Pengue W. Agricultura industrial y transnacionalizaciónen América Latina. "La transgénesis de un continente". Buenos Aires: PNUMA; 2005. 221 p
15. Teubal M. La expansión del modelosojeroen Argentina. De la producción de alimentos a los commodities. RealidadEconómica. 2006;220:71-96
16. Rulli J. Repúblicasunidas de la soja. Realidadessobre la producción de soja en América del Sur. Buenos Aires: GRR; 2008. 294 p
17. Reboratti C. Un mar de soja: la nuevaagriculturaen Argentina y sus consecuencias. Revista de Geografía Norte Grande. 2010;45:63-76
18. Morina JO. Geografías de la agricultura industrial sojeraen la Argentina. "Viaje de ida". Luján: UNLu; 2010. 146 p
19. Gómez Lende S. El modelosojeroen la Argentina (1996-2014), un caso de acumulaciónpordesposesión. Mercator - Revista de Geografia da UFC. 2015;14(3):7-25

20. Teubal M. Soja transgénica y crisis del modeloagroalimentarioargentino. RealidadEconómica. 2003;196:52-74

21. Teubal M, Rodríguez M. Agro y alimentosen la globalización. Una perspectivacrítica. Buenos Aires: Ediciones La Colmena; 2002. 208 p

22. Hernández V. La ruralidadglobalizada y elparadigma de losagronegociosen las pampas gringas. In: Gras C, Hernández V, editors. La Argentina rural. De la agricultura familiar a losagronegocios. Buenos Aires: Biblos; 2009. pp. 39-64

23. Domínguez D, Sabatino P. Con la soja al cuello: crónica de un paíshambriento productor de divisas. In: Alimonda H, editor. Los tormentos de la materia. Aportes para unaecologíapolíticalatinoamericana. Buenos Aires: CLACSO; 2006. pp. 249-274

24. Ribeiro S. Transgénicos: un asalto a la salud y al medio ambiente. RealidadEconómica. 2000;75:70-87

25. Barri F, Wharren J. El modelo del "agronegocio" en la Argentina: elparadigmacientificista-tecnológico. In: Giarracca N, Teubal M, editors. Actividadesextractivasenexpansión: "reprimarización de la economíaargentina". Buenos Aires: Antropofagia; 2013. pp. 73-96

26. Perelmuter T. Bienescomunes vs. mercancías: las semillasendisputa. Un análisis del rol de la propiedadintelectualenlosactualesprocesos de cercamientos. Sociedades rurales, producción y medio ambiente. 2011;11(22):53-86

27. Melón D. El comercio del "oroverde". El patentamiento de la diversidadgenética y la vida. In: Melón D, editor. La Patria sojera. El modeloagrosojeroenel Cono Sur. Buenos Aires: Editorial El Colectivo; 2014. pp. 59-74

28. Altieri MA, Pengue W. La soja transgénicaen Argentina: unamaquinaria de hambre, deforestación y devastación socio-ecológica. Ecología Política Cuadernos de Debate Internacional. 2005;30:87-93

29. Pengue W. Agua virtual, agronegociosojero y cuestioneseconómico-ambientalesfuturas. RealidadEconómica. 2006;223:58-77

30. Pengue W. Suelo virtual, biopolítica del territorio y comerciointernacional. Fronteras. 2010;7:12-25

31. Gómez Lende S. Orden global, "ordenirracional" Agricultura de exportación, devastación de bosques nativos y crisis socio-ambientalen Argentina. Ensayos e InvestigacionesGeográficas. 2009;7(VII):37-52

32. Gómez Lende S, Velázquez G. Orden global, reestructuraciónproductiva y crisis ambiental: análisis de la relación entre deforestación, soja transgénica y silviculturaen la Argentina contemporánea. In: Cacace GP, Gómez ME, Morina JO, Suevo ME, editors. Geografíasregionales y extractivismosen la Argentina de losBicentenarios. Luján: UNLu; 2013. 23-86

33. Rodríguez Striebeck MP. Territoriosendisputa. Conflictosporelacceso y uso de la tierra. In: Melón D, editor. La Patria sojera. El modeloagrosojeroenel Cono Sur. Buenos Aires: Editorial El Colectivo; 2014. pp. 107-121

34. Carrasco A. Glyphosate affects development in Xenopus embryos. In: 68th Meeting of Developmental Biology Society; 23-27 July 2009. San Francisco: Society for Developmental Biology; 2009

35. Domínguez D, Sabatino P. La muerte que vieneenelviento. La problemática de la contaminaciónporefecto de la agriculturatransgénicaen Argentina y Paraguay. In: VVAA, editors. Los señores de la soja. La agriculturatransgénicaen América Latina. Buenos Aires: EdicionesCiccus-CLACSO; 2010. pp. 31-121

36. Verzeñassi D. El modeloagrosojero y suimpactoennuestrasvidas. In: Melón D, editor. La Patria sojera. El modeloagrosojeroenel Cono Sur. Buenos Aires: Editorial El Colectivo; 2014. pp. 31-48

37. Sández F. La Argentina fumigada. Agroquímicos, enfermedad y alimentosen un paísenvenenado. Buenos Aires: Espejo de la Argentina-Planeta; 2016. 455 p

38. Gómez Lende S. Usos del territorio, acumulaciónpordesposesión y derecho a la saluden la Argentina contemporánea: elcaso de la soja transgénica. GEOgraphia – Revista do Programa de Pós-Graduaçãoem Geografia da Universidade Federal Fluminense. 2017;19(39):3-15

39. INDEC. Exportacionessegúncomplejosexportadores. Año 2014. Buenos Aires: Instituto Nacional de Estadística y Censos; 2015. Available from: https://www.google.com.ar/?gws_rd=ssl#q=Exportaciones +seg%C3%BAn+complejos+exportadores.+A%C3%B1o+2014+Indec+excel

40. INDEC. Evolución de puestos de trabajo y promedios de remuneraciones y costosalarial de losasalariadosregistrados. Segundo trimestre 2015. Buenos Aires: Instituto Nacional de Estadística y Censos; 2015. 23 p

41. Anlló G, Bisang R, Salvatierra G. Cambiosestructuralesen las cadenasagropecuarias. De lo primario a las cadenasglobales de valor. Santiago de Chile: CEPAL; 2010. 101 p

42. INDEC. Censo Nacional Agropecuario 2008. Resultadosdefinitivosporprovincias y departamentos. Buenos Aires: Instituto Nacional de Estadística y Censos; 2016. Available from: http://www.indec.gov.ar/nivel4_default.asp?id_tema_1=3&id_tema_2=8&id_tema_3=87

43. MECON. Complejooleaginoso. Serie Producción regional porcomplejos productivos. Buenos Aires: Ministerio de Economía y Finanzas Públicas; 2011. 28 p

44. Silveira ML. Um país, umaregião. Fim de século e modernidadesna Argentina. São Paulo: LABOPLAN-USP; 1999. 488 p

45. Morina JO, Cacace GP. Capitalismoagrario y expansiónsojeraen la Argentina: "un extractivismo sin retorno" In: Cacace GP, Gómez ME, Morina JO, Suevo ME, editors. Geografíasregionales y extractivismosen la Argentina de losBicentenarios. Luján: UNLu; 2013. pp. 287-328

46. INDEC. Censo Nacional Agropecuario 1988. Resultadosdefinitivosporprovincias y departamentos. Buenos Aires: Instituto Nacional de Estadística y Censos; 1995. 425 p

47. Gras C. Expansiónagrícola y agriculturaempresarial. El casoargentino. Revista de CienciasSociales. 2013;26(32):73-92

48. CIARA-CEC. Estadísticasnacionales. Buenos Aires: Cámara de la IndustriaAceitera de la República Argentina; 2016. Available from: http://www.ciara.com.ar/estadisticasNac.php

49. Toledo López V. Los agrocombustiblescomo un eje del extractivismoen la Argentina. In: GiarraccaN, TeubalM, editors. Actividadesextractivasenexpansión: "reprimarización de la economíaargentina" Buenos Aires: Antropofagia; 2013. pp. 137-158

50. CAC. Perfilesexportadoresprovinciales 2015. Buenos Aires: Cámara Argentina de Comercio; 2015. 58 p

51. Teubal M. Soja y agronegociosen la Argentina: la crisis del modelo. Lavboratorio. Estudiossobrecambioestructural y desigualdad social. 2008;10(22):5-7

52. Klippahn A, Enz D, Tierras SA. Crónicas de un paísrematado. Buenos Aires: Aguilar; 2006. 254 p

53. Giarracca N, Palmisano T. Tres lógicas de producción de alimentos: "Hay alternativas al agronegocio" In: Giarracca N, Teubal M, editors. Actividadesextractivasenexpansión: "reprimarizacióndelaeconomíaargentina" Buenos Aires: Antropofagia; 2013. pp. 19-44

54. MINAGRI. Relevamiento y sistematización de problemas de tierra de losagricultoresfamiliaresen la Argentina. Buenos Aires: Ministerio de Agricultura, Ganadería, Pesca y Alimentación; 2013. 96 p

55. REDAF. Conflictossobretenencia de tierra y ambientalesen la región del Chaco argentino 3° informe. Reconquista: Red Agroforestal Chaco Argentina, Observatorio de Recursos Naturales, Tierras y Medioambiente; 2013. 100 p

56. Naharro N, Álvarez A. Estudio de caso: acaparamiento de tierras y producción de soja enterritoriowichí. Salta-Argentina. Salta: Brot für die Welt-Asociana; 2011. 39 p

57. Aranda D. Tierra arrasada. Petróleo, soja, pasteras y megaminería. Radiografía de la Argentina del Siglo XXI. Buenos Aires: Sudamericana; 2015. 430 p

58. Delgado O. La ruta de la soja enelNoroeste Argentino. In: Rulli J, editor. Repúblicasunidas de la soja. Realidadessobre la producción de soja en América del Sur. Buenos Aires: GRR; 2008. pp. 132-158

59. SAyDS. Monitoreos de la superficie de bosque nativo de la República Argentina. Períodos 1998-2002, 2002-2006, 2006-2011, 2011-2013 y 2013-2014. Regiones forestales Parque Chaqueño, Yungas Selva Paranaense y Espinal. Buenos Aires: Secretaría de Ambiente y Desarrollo Sustentable de la Nación; 2007-2015. 772 p

60. FAO. Situación de los bosques del mundo 2007. Roma: Organización de las NacionesUnidas para la Agricultura y la Alimentación; 2007. 157 p

61. FAO. El estado de los bosques del mundo 2012. Roma: Organización de las NacionesUnidas para la Agricultura y la Alimentación; 2012. 51 p

62. Vallejos M, Volante JN, Mosciaro MJ, Vale LM, Bustamante ML, Paruelo JM. Transformation dynamics of the natural cover in the Dry Chaco ecoregión: A plot level geo-database from 1976 to 2012. Journal of Arid Environments. 2015;123:3-11

63. REDUAS. Informe 1° Encuentro Nacional de Médicos de Pueblos Fumigados. Córdoba: Facultad de CienciasMédicas, Universidad Nacional de Córdoba; 2013. 40 p

64. DPN. Niñez y riesgoambientalen la Argentina. Buenos Aires: Defensor del Pueblo de la Nación; 2009. 77 p

65. GRR. Pueblos fumigados. Los efectos de losplaguicidasen las regiones sojeras. Buenos Aires: Grupo de Reflexión Rural, Editorial del Nuevo Extremo; 2009. 254 p

66. Ribeiro S. Monsanto y la soya argentina. La Jornada; 2005. Available from: http://www.etcgroup. org/es/content/monsanto-y-la-soya-argentina

67. CONICET-MSN. Informe-Evaluación de la informacióncientíficavinculada al glifosatoensuincidenciasobre la saludhumana y elambiente. Buenos Aires: Consejo CientíficoInterdisciplinario, CONICET-Ministerio de Salud de la Nación; 2009

68. Bertello F. Para elGobierno, se terminóelconflicto con Monsanto. Diario La Nación; 2016. Available from: http://www.lanacion.com.ar/1911960-para-el-gobierno-se-termino-el-conflicto-con-monsanto

08 MAIN DRIVING FORCES AND CHALLENGES OF AGRICULTURAL MARKET INTEGRATION IN THE COMMONWEALTH OF INDEPENDENT STATES

1. INTRODUCTION: AN OVERVIEW

The transition in market structures within the Commonwealth of Independent States (CIS) from centrally planned economies to open markets has led to privatization and consolidation of supply chain actors, particularly in the processing and retailing sectors. This shift has sparked significant interest among consumers, supply chain members, and policy makers in understanding how the market operates.

The Former Soviet Union/CIS countries' trade relations were significantly impacted by recent agricultural price fluctuations, geopolitical events such as the creation and expansion of the Eurasian Economic Union, and the Russian agricultural import ban. These factors also had notable effects on domestic food prices. These advancements have made food pricing a priority on political agendas globally, particularly in CIS countries. Although a significant rise in commodity prices typically leads to higher consumer costs, declines in commodity prices may not immediately result in lower consumer prices. Therefore, comprehending the price transmission mechanisms within the supply chain is essential for establishing an appropriate agricultural policy that enables most market participants to benefit from a sustainable distribution of value added [1].

Agricultural supply networks connect agricultural farmers with food consumers through the food processing industry and food merchants. This study centers on the supply chain phase of agricultural production carried out by farmers. We examine market integration and efficiency in agricultural supply chains in the CIS countries using a price transmission technique. How much do price shocks spread between regions and from the global market to domestic markets? Efficient markets with good integration can help mitigate price increases caused by regional production shortages and prevent prices from exceeding global market levels.

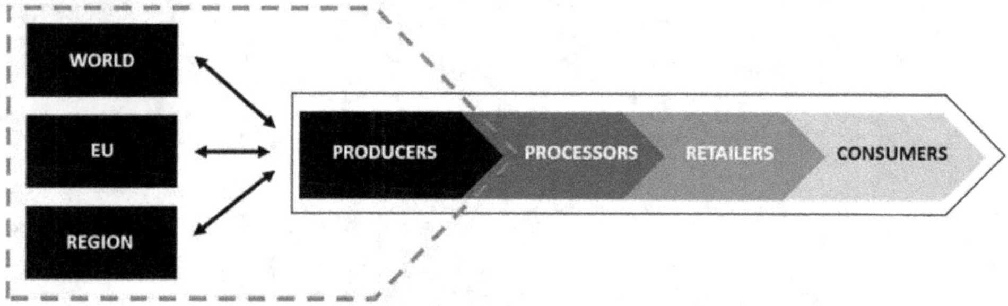

Figure 1. Research scope. Source: own illustration.

Generally, a unique concept of market integration and efficiency does not exist [2]. This chapter operates under the assumption that a well-operating market is spatially efficient and exhibits substantial market integration. Therefore, price fluctuations in one area are rapidly transferred to other regions, leading to interregional trade movements when price disparities surpass trade expenses[2]. Regional prices vary mainly due to trade expenses between areas (Law of One Price), therefore successful trade arbitrage opportunities are not sustainable. An efficient market is defined by reasonable trade costs influenced by various factors such as proximity to other markets, the standard and capacity of transportation and communication facilities, corruption levels, market risks, and regulatory obstacles. phytosanitary license and inspection requirements [3].

We study agricultural supply chains in former Soviet Union countries that are connected to adjacent countries or the global market[1] through agricultural imports and exports. We use a comparative technique to analyze five agricultural commodities (wheat, pork, beef, poultry, and milk) across eight nations in the CIS region (Armenia, Azerbaijan, Belarus, Georgia, Kazakhstan, Moldova, Russia, and Ukraine). We use a standardized approach to analyze a consistent sample of data with the same frequency and analysis period for the purpose of ensuring comparability. There is a scarcity of data on agricultural pricing in the CIS countries, leading to a research gap. Exceptions are references [4] on Georgian bread, sugar, beef, and pork markets, and reference [5] focusing on the wheat flour market in Georgia.

In the rest of this chapter, we will outline our methodological approach and the data utilized in our analysis in Section 2. Section 3 describes the agricultural trade features of the CIS countries. Section 4 presents empirical findings, while Section 5 discusses the main factors influencing market integration. Section 6 contains the conclusions.

2. METHODOLOGICAL APPROACH AND DATA

We analyze spatial transmission of different agricultural and food prices between various CIS countries and between CIS countries and the world market. Specifically, we focus on wheat, meat (beef, pork, and poultry), and milk in eight CIS countries in the region. An overview on the countries included in our analysis is presented in Appendix.

We analyze on the integration of agricultural markets in CIS countries with their regional trade partners within the CIS region but also with the EU market and the world market. We evaluate the integration of domestic CIS markets with the export markets in other CIS countries within the region. We also examine how price changes from the EU market and international markets are transmitted to the domestic markets in the CIS region.

Focusing on the bivariate price transmission model, we characterize long-run dynamics between price pairs consisting of a domestic price of one of the CIS countries and a world market price (or price of some CIS exporting country).[2] We refer to a domestic market as fully integrated in the world market, if price changes on the world market are completely transmitted to the domestic market. A long-run price equilibrium is given as

$$P_t^d = \alpha + \beta P_t^w + \varepsilon_t \qquad (1)$$

where P_t^d and P_t^w are natural logarithm of domestic and world market prices at time tespectively,the intercept parameter α,the slope parameter β, and the residual ε t denoting the stationary disturbance term. The long-run price transmission parameter β indicates the degree (in %) to which price changes on the world market are transmitted to the domestic market. Perfect market integration is given if β = 1, implying that a 1% price change on the world market is transmitted by 100% to the domestic market, leading to a 1% price change on the domestic market. We apply the augmented Dickey-Fuller (ADF) test [6] to test on stationarity of our data series. We test for cointegration between the non-stationary price series using the Johansen test of linear cointegration [7]. In case, we find domestic and world market prices cointegrated, the Ordinary least squares (OLS) regression yields consistent and efficient estimates of the long-run equilibrium parameters [8].

As already mentioned above, data availability is one of the most critical issues when it comes to research on food markets in the CIS countries. In this study, we have created a unique data set for selected food products by combining different data sources, ranging from statistical offices of the respective CIS countries, via different country reports published by various international institutions (e.g., WB, FAO, and OECD), all the way to expert interviews. According to data availability, we are able to conduct the analysis for wheat, meat (pork, beef, and poultry), and milk (i.e., milk powder) markets of all eight selected CIS countries.

Concerning wheat prices, we use producer, import, and export prices of wheat in different countries. Data are sorted into two groups. The first group refers to Armenia, Azerbaijan, Georgia, Ukraine, and Moldova where we use average producer wheat prices. Since domestic producer prices for wheat in Georgia are not available, we use the cost, insurance, and freight (i.e., CIF) import price. The second group consists of the Customs Union members, i.e., Russia, Kazakhstan, and Belarus. Considering that Russia and Kazakhstan are, besides Ukraine, the main wheat exporters not only to the CIS region but also to the world market as well, in addition to average producer prices, we also account for their free on board (i.e., FOB) export prices observed at the Black Sea harbors. For Belarus, we use the governmentally fixed wheat purchase prices. We use average French FOB price at the port of Rouen (representative price for the Marché à Terme International de France (MATIF) commodity futures market in Europe) as a measure for the EU market price. Average FOB soft red winter wheat prices of the USA serve as representative for the world market price. The time period covered by our data set ranges from 2006 until 2014 and differs between countries and products. Thus, the number of observations varies between 74 (for Moldova) and 130 (for Belarus).

Our data set comprises pork prices for Armenia, Georgia, Belarus, Kazakhstan, Russia, and Ukraine. In addition, we consider pork prices for the EU, Brazil, and USA, which are some of the most important pork exporters to the CIS countries. Denmark, Germany, and Spain together have exported more than 50% of total EU pork export to the CIS in 2013. Our data set contains two types of pork prices. First, for Ukraine, we account for producer prices, which refer to the price paid for 1 kg of live animal. Second, we account for the producer price for 1 kg of deadweight for all other countries. The conversion of the prices between 1 kg of live animal and deadweight is done by using the conversion rate of 0.59 for Ukraine [9]. Our data set covers the period from January 2004 until December 2014. The number of observations differs between countries, from 45 observations for Ukraine to 132 observations for Armenia, Russia, EU, and USA.

As was the case for pork, we account for beef prices for Armenia, Georgia, Russia, Kazakhstan, Belarus, and Ukraine. For the EU, we use the EU average beef price. For the world price, we focus on beef prices of USA, Brazil, Uruguay, Argentina, and Australia, which are directly involved in beef export to CIS countries. Our data set covers the period from January 2004 until December 2014. The number of observations differs between countries, from 45 observations for Ukraine to 132 observations for Armenia, Russia, EU, and USA.

Due to very limited data availability for poultry, only Georgia as a representative for the Caucasian countries is covered by our analysis. For the EU price, we account for both EU average and country-specific average poultry prices. Here, we selected Netherlands and Germany as two of the most important EU exporters of poultry meat to CIS countries. The data set covers a period from January 2004 until December 2014. The number of observations differs between countries, from 45 observations for Ukraine to 132 observations for Russia.

The milk prices used for the analysis are average monthly milk producer prices. For the selected international markets, we use EU whole milk powder (WMP) export prices and FOB Oceania WMP prices. In addition, we use domestic WMP prices for Netherlands and Germany, countries that are the largest EU WMP exporters to the CIS countries accounting for 21 and 16% of the total CIS import, respectively. To compare the fluid milk producer prices of CIS countries with the WMP prices of the selected international markets, we transform the fluid milk prices into WMP prices using conversion factors obtained by the national experts of the selected CIS countries [10]. Thus, for Armenia, we use a conversion factor of 7.5, for Belarus 7.3, for Kazakhstan 7.5, For Russia 7, and for Ukraine 8.3. Our data set covers a period from January 2004 to December 2013. As in previous cases, the number of observations differs between the countries, ranging from 36 observations for Ukraine to 106 observations for most other countries.

Overall, considering that most of the agricultural trade of the CIS countries is done in US Dollars (USD), prices in domestic currencies are converted in USD. Thus, all prices used for the analysis refer to USD per measurement unit.

3. AGRICULTURAL TRADE OF THE CIS COUNTRIES

In the following, we provide an overview on agricultural trade of the CIS trade regarding wheat, pork, beef, poultry, and milk which provides the basis for the interpretation of our empirical results in Section 4.

Wheat production and trade are of a great importance for the CIS countries. From one side, Russia, Ukraine, and Kazakhstan are becoming large wheat exporters that are important not only for the CIS region but also for the international markets. This is especially the case for Russia that became the largest wheat exporter in the world in 2016 of about 40 million ton. On the other side, Caucasian countries (i.e., Armenia, Azerbaijan, and Georgia) import more than 90% of wheat from the CIS region, notably from Russia and Kazakhstan [11]. From the regional perspective, Russian wheat is the most important for domestic consumption, and its quality is usually improved by imports of the high-quality Kazakh wheat. This is mainly the case for Caucasian countries. Nevertheless, Kazakh wheat is almost the only source of wheat for some Central Asian countries (i.e., Uzbekistan, Tajikistan, and Kyrgyzstan). Besides mentioned wheat exporting and importing CIS countries, Belarus and Moldova produce a sufficient amount of wheat for domestic production but are not large wheat exporters.

All the selected CIS countries are net pork importers. The EU (27) is the largest supplier of pork to the CIS countries with a share of 56% (in 2013). Only about 2% of pork is traded regionally, where the main exporters are Moldova, Belarus, Russia, and Kazakhstan. Significant market changes are recorded for the members of the Customs Union, where each member country is supposed to adjust their meat import tariff according to the Union's common trade policy. This was particularly important for Kazakhstan, which did not have any significant internal market protection before [12]. Nevertheless, most of the selected CIS countries have a higher level of price protection for pork compared to other products [13].

In comparison with pork, CIS beef markets are considerably different. Namely, Caucasian countries (i.e., Armenia, Azerbaijan, and Georgia) are net beef importers. The same is true for Russia and Kazakhstan. About 89% of beef is imported from some of the international markets (excluding EU (27)). Only about 8% is imported from some of the CIS countries. On the other hand, the main beef exporting CIS countries are Belarus, Moldova, and Ukraine. Overall, beef is one of the products that have the smallest level of price protection among CIS countries [13].

Similar to the CIS beef markets, some of the CIS countries are net importers, and some are net poultry exporters. Concerning imports, about 17% is imported from the EU (27) markets, 22% from other CIS countries, and 61% from other international markets (in 2013, [14]). The net importing countries are Armenia, Azerbaijan, Georgia, Moldova, Russia, and Kazakhstan. On the other side, some of the net importers also engage in poultry export but mainly on the regional markets. The only two net exporting countries are Belarus and Ukraine. As is the case with pork, poultry also has high level of price protection among CIS countries [13].

Almost half of the total dairy products imported to the selected CIS countries originate from the region. Namely, 47% of total diary imports originate from some of the CIS countries, where Belarus, Russia, and Ukraine are the largest regional exporters. On the other hand, about 44% of the total CIS dairy imports originate from the EU, where Netherlands, Germany, and Finland represent the largest exporters, covering 50% of the EU exports to CIS countries. Only about 9% of dairy products are imported from some of the world markets (excluding CIS and EU). Concerning milk, it is mainly imported in the form of a milk powder. About 65% of total milk powder imports by the CIS countries originate from Belarus and Ukraine, which are net exporters. Further, about 23% of milk powder is imported from EU (27) markets.

4. EMPIRICAL RESULTS

Empirical results are presented for each of the five agricultural products individually. Besides the comparison of average price levels between countries, we provide a detailed overview on the pattern of market integration for each product. Market integration for the different agricultural products is analyzed on an aggregated level for the CIS countries on average and on a disaggregated level for individual countries. Market integration of the CIS countries is investigated regarding their regional integration in the CIS, the integration with the EU market and the world market.

4.1. Wheat

Domestic price and trade policies are among most important factors determining domestic price developments. Due to different production and trade characteristics of the CIS countries covered in this study, domestic price developments differ between the selected countries (Figure 2). Thus, wheat prices in Armenia, Azerbaijan, and Georgia, three wheat importing countries, are higher compared to the CIS average. On the other hand, wheat prices in Russia and Ukraine, two large wheat exporters, are on average lower compared to the average CIS prices and selected international prices. The most extreme case is Belarus, where the domestic government keeps wheat prices at an artificially low level to support domestic meat production [15]. Furthermore, wheat producer prices in the whole CIS region were affected by wheat export restrictions imposed by Russia, Ukraine, and Kazakhstan between 2006 and 2012 [16], which ultimately had a strong effect on consumers and overall food security in these countries.

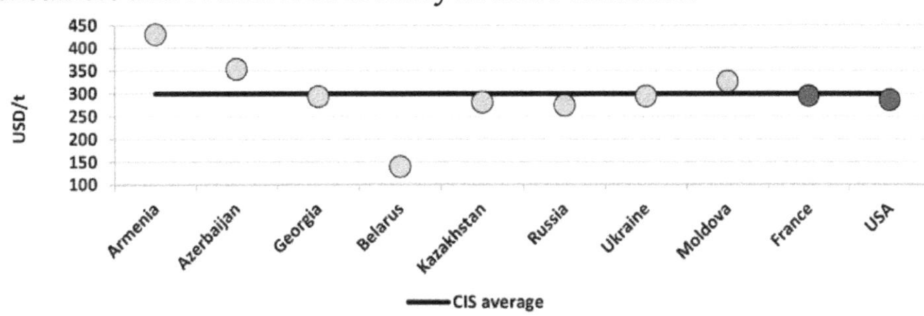

Figure 2. Average wheat price level in the CIS countries and selected international markets (base: 2011–2013). Source: Statistical offices of the respective CIS countries [17, 18], own illustration.

The analysis of market integration indicates that the CIS wheat markets are strongly integrated within the region (integration between CIS countries) and with the EU (i.e., France) and world markets (i.e., USA; Figure 3). Surprisingly, integration in world wheat markets is highest amounting to about 76%. Regional wheat market integration lies by on average 68%, and integration with the EU wheat market lies by 61%.

Figure 4 provides a detailed overview on the pattern of market integration. On the regional level, Georgia and Moldova are strongest integrated with the other CIS countries (Figure 4). Moderate regional integration is recorded for Armenia and Azerbaijan, countries that are heavily dependent on wheat imports from other CIS countries. A moderate level of integration might indicate certain market inefficiencies, which might be connected to certain market regulations set by the government, for example, in the case of Azerbaijan.

The integration of the CIS countries within the CIS regional market is very similar to their integration in the EU market and the world market. The main reason is that both markets, EU and world, are used as reference markets for price negotiations in wheat trade by CIS traders. Price discounts or mark-ups might result from quality differences and internal price determining factors. It is striking that the Ukrainian wheat market has stronger integration with world markets compared to the EU. Overall, the Russian, Ukrainian, and Moldavian wheat markets are strongly integrated with both the EU and world markets. These results are in accordance with the trade status of these countries. Namely, Russia and Ukraine are large wheat exporters not only in the CIS region but also on the international level as well. Thus, international price changes have a great impact on their domestic producer prices.

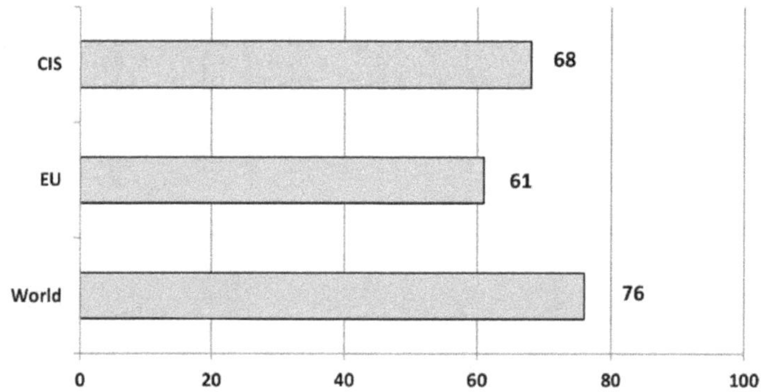

Figure 3. CIS wheat market integration—aggregated level. Source: Own calculation based on the estimation results, own illustration.

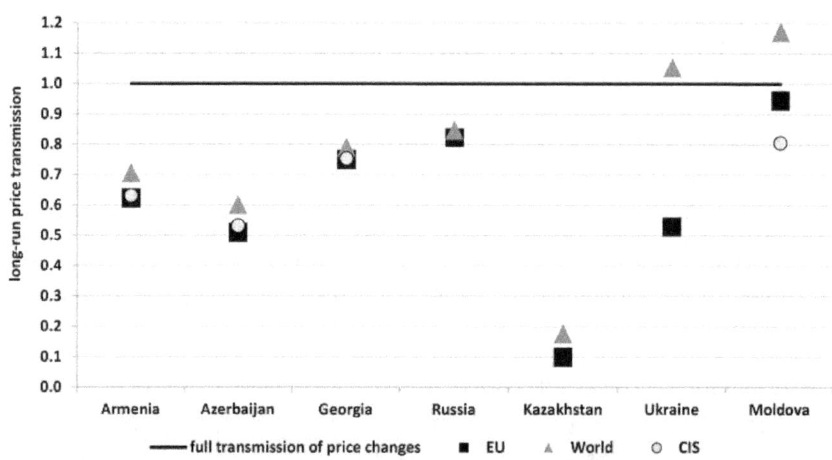

Figure 4. CIS wheat market integration—disaggregated level. Source: Own calculation based on the results, own illustration.

On the other hand, Moldova is also an exporting country but with no significant relevance for the regional and international wheat trade. It is interesting to observe that there is almost no integration between Kazakh and international wheat markets. The main reason is that Kazakh wheat export is mainly concentrated on Central and East Asian countries, and thus, Kazakh domestic prices do not react much to price shocks from the "Western" markets.

4.2. Pork

The 3-year pork price averages presented in Figure 5 indicate that the CIS prices are higher than the prices on the EU (i.e., Denmark, Germany, and Spain) and world markets (i.e., Brazil and USA). For the Customs Union members (i.e., Belarus, Russia, and Kazakhstan), we can observe that Russian pork prices are at the level of the CIS average price but by 20 and 11% higher than the average Belarussian and Kazakh prices, respectively. The Belarus price level is not reflecting actual market conditions but is rather the result of the price targeting by the Belarussian government [13].

The results of the price transmission analysis indicate that the CIS countries are on average well integrated with both regional and international pork markets. The highest integration level is recorded with the EU market where we observe full transmission of price shocks (Figure 6). Besides the EU, regional markets seem to be very well integrated among each other, where about 80% of the price shocks are transmitted in the long run. In particular, regional market integration is very strong between Belarus, Russia, and Kazakhstan. International market price shocks are only by about 50% transmitted to the domestic markets in the CIS countries. Obtained results are in accordance with the CIS pork trade developments where we observe that 56% of the total CIS pork import (for selected CIS countries) originates from the EU.

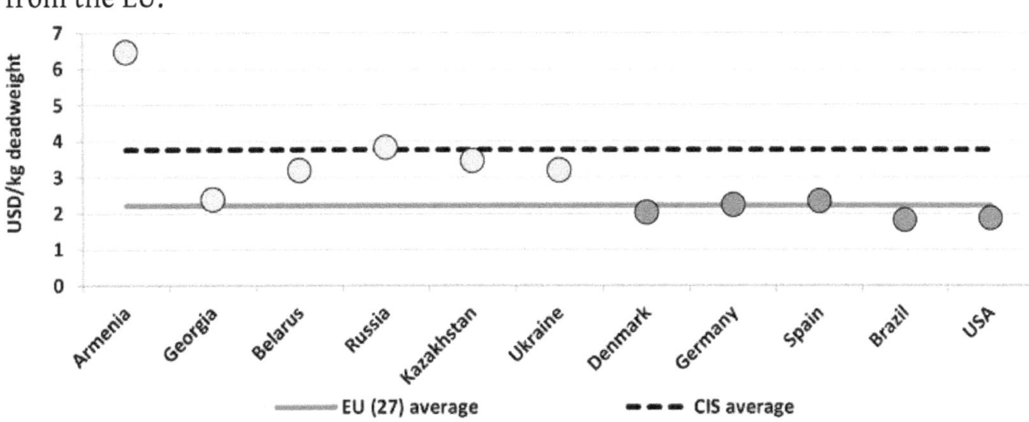

Figure 5. Average pork price level in the CIS countries and selected international markets (base: 2011–2013). Source: Statistical offices of the respective CIS countries and [19], own illustration.

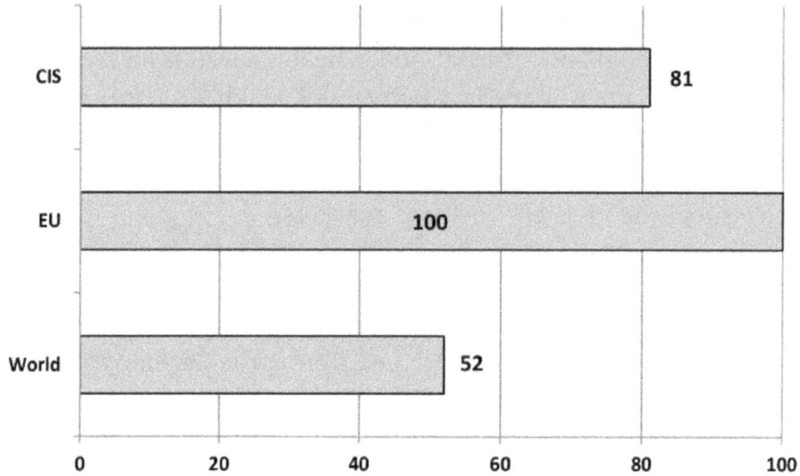

Figure 6. CIS pork market integration—aggregated level. Source: Own calculation based on the estimation results, own illustration.

A cross-country comparison of the price transmission results is presented in Figure 7. As already mentioned, our results indicate highest integration of the CIS pork market with the EU. This is certainly the case for Belarus, Russia and Kazakhstan, members of the Customs Union, for which we found full transmission of price shocks from the EU market in the long run. These results are not surprising considering that Belarus, Russia, and Kazakhstan import 98, 59, and 57%, respectively, of their total pork import from the EU. On the other hand, Belarus, Russia, and Kazakhstan have a very low level of integration with other CIS countries. Concerning market integration with the international markets (Figure 7), our results indicate strong integration of Georgian and Armenian pork markets, while Belarus, Russia, and Kazakhstan show very weak integration.

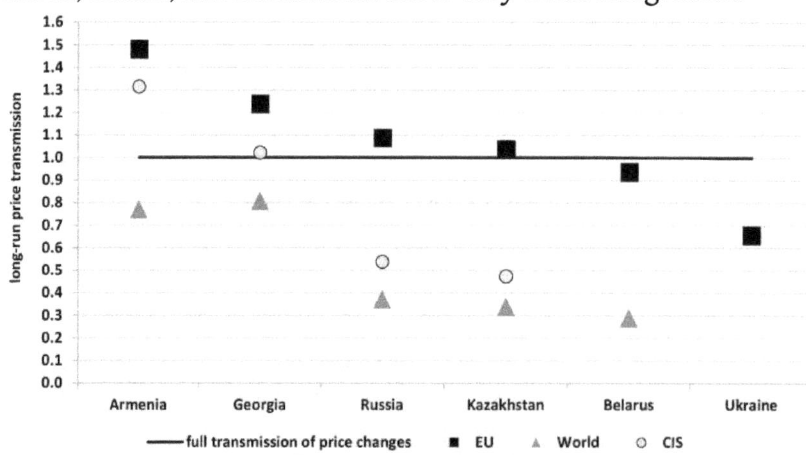

Figure 7. CIS pork market integration—disaggregated level. Note: Missing marker means no integration with the respective market. Source: Own calculation based on the results, own illustration.

4.3. Beef

The 3-year beef price averages presented in Figure 8 indicate that the average CIS prices are 29% lower than the EU price (i.e., EU average price) but are similar to the price level of the world markets (i.e., Brazil, Uruguay, USA, Argentina, and Australia). Armenia, Russia, and Kazakhstan have the highest beef prices, which are 50, 35, and 20% higher than the CIS average prices, respectively. Our data indicate that Ukraine has relatively low beef prices compared to other CIS countries, the EU and the world market, with the price level about 60% lower than the CIS average. Overall, the data presented in Figure 8 show significant differences in average beef prices between the CIS countries.

Our price transmission results indicate very strong integration of CIS beef markets both regionally and internationally. We find that price shocks are perfectly transmitted from the EU market and by 88% from the world market to the CIS countries on average. But even regional integration is high amounting to 86% of the CIS on average (Figure 9).

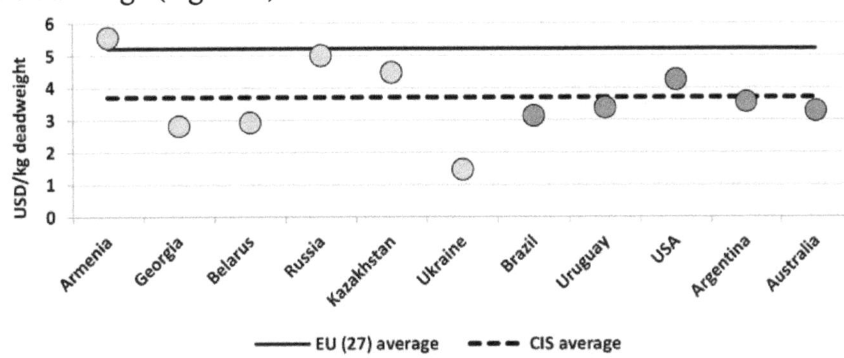

Figure 8. Average beef price level in the CIS countries and selected international markets (base: 2011–2013). Source: Statistical offices of the respective CIS countries and [19], own illustration.

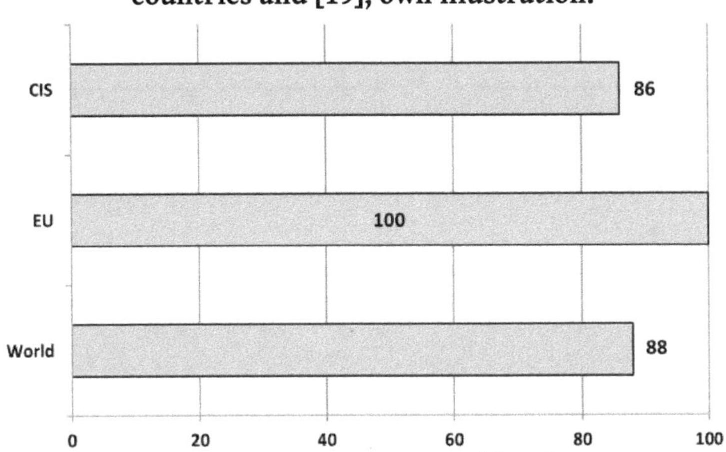

Figure 9.CIS beef market integration—aggregated level. Source: Own calculation based on the estimation results, own illustration.

The cross-country market integration results on the disaggregated level presented in Figure 10 indicate that Kazakhstan and Belarus are strongly integrated with other regional markets. A moderate level of regional integration is recorded for Russia, while Georgia seems not to be integrated with other regional markets in the long run. Concerning market integration with the EU, our results indicate almost full transmission of price shocks in the long run. In contrast, our results do suggest no market integration of the Ukrainian market with the EU.

4.4. Poultry

The 3-year average poultry prices presented in Figure 11 indicate that the CIS prices are 26% lower than the EU price (i.e., Netherlands and Germany) and are similar to the world market price level (i.e., Brazil and USA). This is especially the case for Georgia, Ukraine, and Kazakhstan, where prices are below the CIS average. In contrast to pork and beef prices, poultry prices are not characterized by a great dispersion for the observed period.

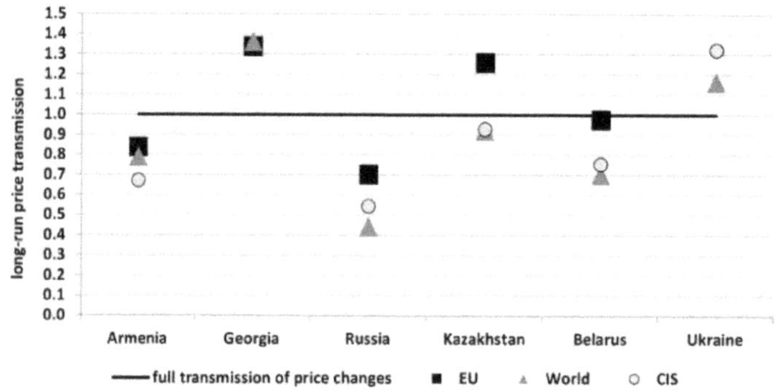

Figure 10. CIS beef market integration—disaggregated level. Note: Missing marker means no integration with the respective market. Source: Own calculation based on the results, own illustration.

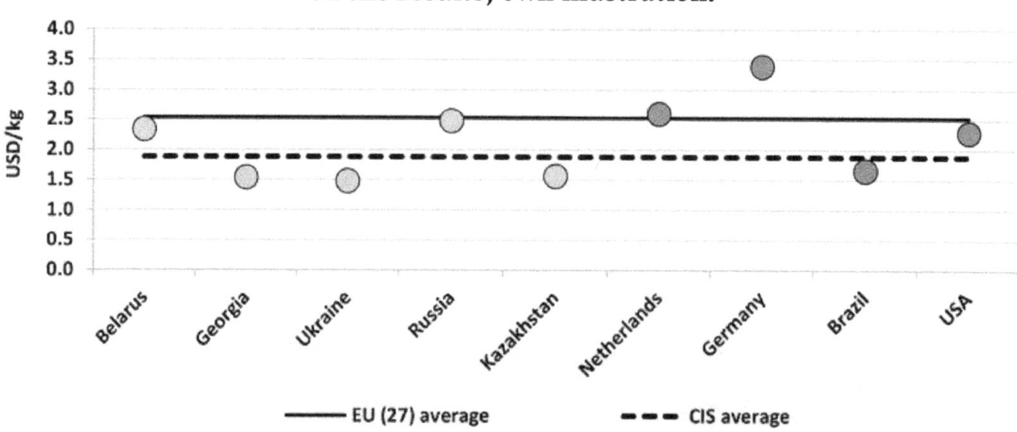

Figure 11.Average poultry price level in the CIS countries and selected international markets (base: 2011–2013). Source: Statistical offices of the respective CIS countries, own illustration.

Concerning the price transmission analysis, our results indicate that the CIS poultry markets are integrated regionally, with the EU and with selected international poultry markets. Strongest integration is recorded with the EU market, where 73% of the price shocks are transmitted to the domestic CIS markets in the long run (Figure 12). Moderate transmission is recorded from regional markets, while relatively slow transmission is recorded from the selected international markets.

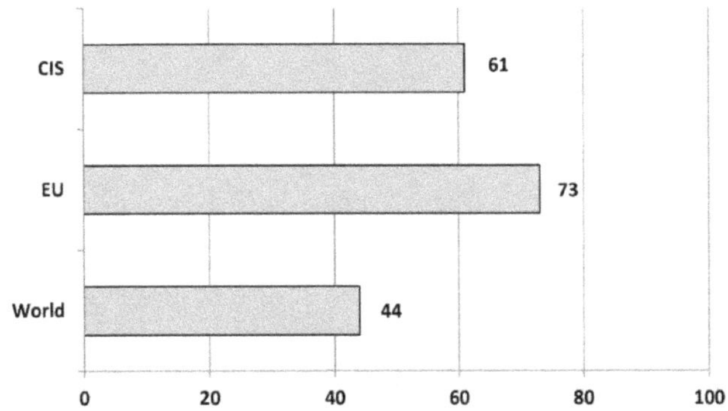

Figure 12. CIS poultry market integration—aggregated level. Source: Own calculation based on the estimation results, own illustration.

The results from the disaggregated cross-country comparison indicate that the Russian poultry prices are strongly integrated with both regional and EU poultry markets, where we observe almost full transmission of price shocks (Figure 13). However, our results of Ukraine indicate no market integration with regional and EU poultry markets.

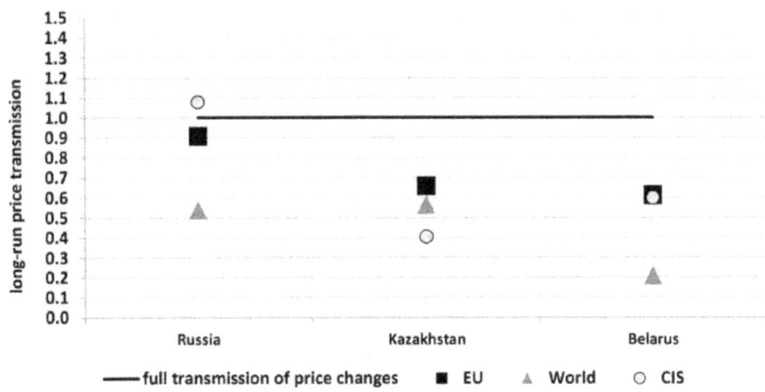

Figure 13. CIS poultry market integration—disaggregated level. Source: Own calculation based on the results, own illustration.

4.5. Milk

For investigating the country-level assessment of CIS milk market integration, we focus on the milk powder prices [i.e., whole milk powder (WMP)] rather than on fluid milk prices.

Figure 14 shows the estimated 3-year WMP price averages for the CIS countries. Compared to selected EU markets (i.e., EU 27 average price, Netherlands, and Germany) and world markets (i.e., Oceania), estimated CIS average prices are by on average 28% lower. The highest WMP prices are recorded for Russia and Kazakhstan (14% above the average), while the lowest prices are recorded for Belarus (27% below the average). Armenian and Ukrainian WMP prices are almost at the level of average CIS WMP prices.

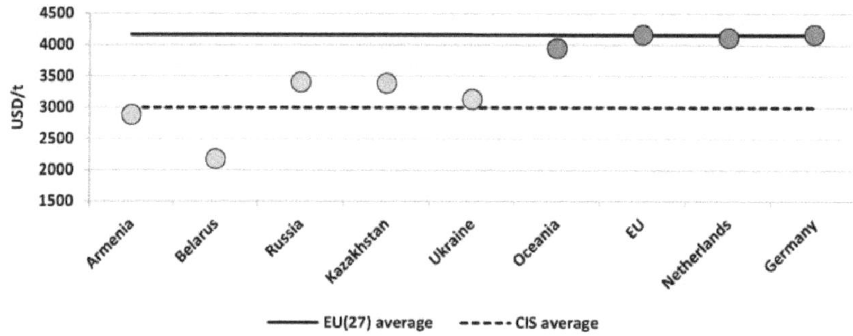

Figure 14.Average milk price level in the CIS countries and selected international markets (base: 2011–2013). Source: Statistical offices of the respective CIS countries and [18], own illustration.

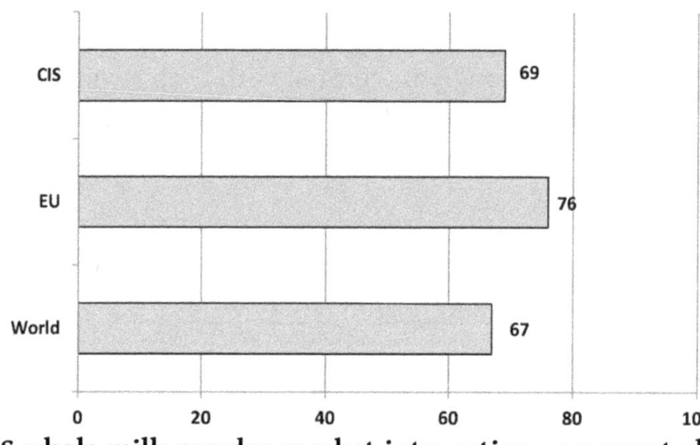

Figure 15. CIS whole milk powder market integration—aggregated level. Source: Own calculation based on the estimation results, own illustration.

The price transmission results indicate relatively moderate integration of the CIS WMP markets with both regional and selected international markets (Figure 15). The results from the cross-country comparison indicate that the Kazakh WMP market is strongly integrated with the regional markets, where we observe almost full transmission of price shocks (Figure 16). The Russian WMP market is strongly integrated with the EU market where we also observe almost full transmission of price shocks. Changes from the selected international market (i.e., Oceania) are transmitted to the Russian and Kazakh markets to the higher degree compared to other CIS countries (Figure 16).

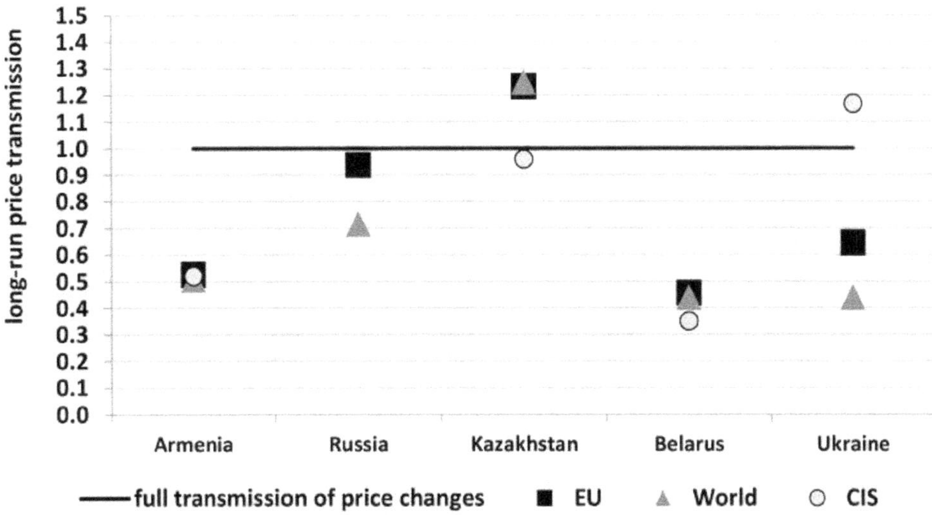

Figure 16. CIS whole milk powder market integration—disaggregated level. Note: Missing marker means no integration with the respective market. Source: Own calculation based on the results, own illustration.

4.6. Summary of Market Integration Results

The results of the price transmission analysis (Figure 17) indicate strongest market integration of the CIS beef market with all three reference markets, i.e., regional CIS, EU, and the world market. As already mentioned, one of the main reasons for almost full transmission of price shocks from reference markets to the domestic CIS markets might be the fact that beef markets in CIS countries have lowest level of price protection compared to other products [13].

The observed strong market integration of the CIS pork markets with the EU market might result from comprehensive pork imports (trade flows) from the EU (until 2014). Similarly, significant pork trade between Belarus, Russia, and Kazakhstan, supported by the removal of trade barriers within the Customs Union, contributed to the rather strong regional pork market integration of the CIS countries.

The strong integration of the CIS wheat markets with the world market is might be explained by two main factors: (1) Russia and Ukraine have advanced to global leaders in wheat trade and thus have a strong influence on world wheat prices; (2) World market prices are used as a benchmark price for regional wheat trade among CIS countries.

Considering that none of the CIS countries is a large poultry and whole milk powder importer or exporter and thus not a significant player on the world market, market integration is lower with the world market compared to EU and CIS regional markets. The EU market prices are used as a benchmark for negotiating poultry and whole milk powder trade between CIS countries.

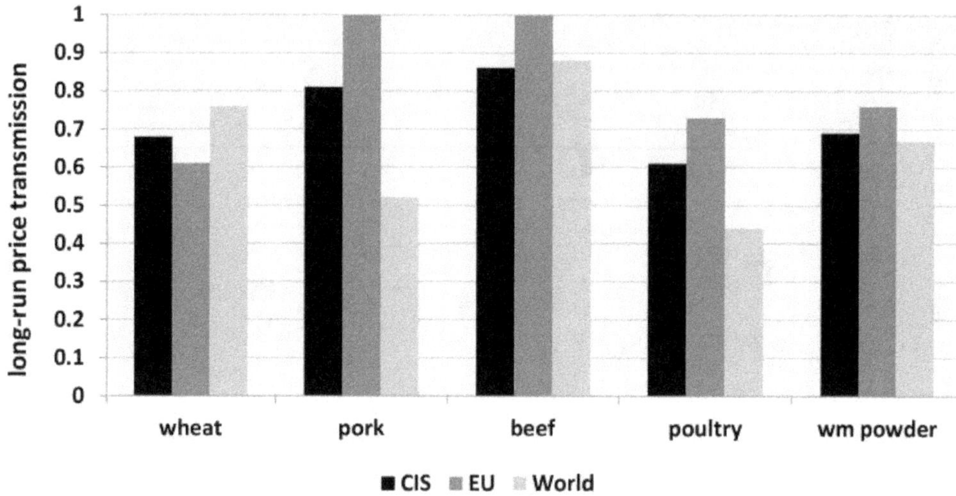

Figure 17. CIS market integration by product. Source: Own calculation and illustration.

Overall, all selected CIS markets are well integrated with the EU market due to market proximity and significant trade flows supported by numerous bilateral trade agreements. Integration with world markets is strong for products where CIS markets are large exporters (e.g., for wheat) or importers (e.g., for beef). Regional market integration greatly depends on trade volumes, infrastructure issues, and market barriers (i.e., removal of tariffs within the Customs Union).

5. DETERMINANTS OF THE CIS MARKET INTEGRATION

Differences in market integration observed between regions or countries might result from the influence of a variety of different factors. In the following, we link major influencing factors, in particular, the level of domestic market support, trade policies, the size of physical trade flows, infrastructure, and bilateral or multilateral trade agreements with our results on market integration of the CIS countries.

First, market support and trade-oriented policy measures have a strong impact on market integration [20]. For markets characterized by a high level of state support, the results indicate moderate or even no integration with international markets. This is especially the case with the CIS pork and poultry markets. On the country level, Belorussian attempts to support livestock production by strongly regulating grain prices is one of the main reasons why the Belorussian wheat market is not integrated with any of the reference markets. On the contrary, CIS markets with almost no state support, such as wheat and beef markets, are strongly integrated with international markets. Besides direct state support, sudden changes in CIS trade policies are significantly affecting CIS market integration. In particular, numerous export-oriented measures (e.g., export bans, export taxes, or export quotas) were implemented on CIS wheat markets in the last decade. The results for Russia and Ukraine indicate that wheat export restrictions significantly decrease the level of market integration and thus the transmission of price changes from international reference markets to domestic markets [21]. Furthermore, the decrease in price transmission is higher for regions within a country that are strongly integrated with international markets when trade is freely possible (e.g., North Caucasus for Russia and Odessa for Ukraine).

Second, the size of physical trade flows and the importance of the reference market for global trade (e.g., US wheat price) play an important role for market integration of the CIS countries. Strong regional integration of the CIS countries is based on the fact that Armenia, Georgia, and Azerbaijan are among the largest CIS wheat importers with wheat imports almost completely originating from other CIS countries (i.e., Russia, Ukraine, and Kazakhstan). At the same time, these countries have strong integration with international markets as well. We argue that price information coming from the main international markets is used by regional CIS traders as a benchmark for negotiating prices in regional trade. Similar considerations could be used for explaining strong market integration between CIS pork markets with regional and EU markets.

Third, underdeveloped infrastructure presents a great obstacle for the market integration of many CIS countries. This is especially visible on a disaggregated country-level analysis. Difficulties to obtain sufficient railway wagons, small capacity of port terminals on the Caspian Sea, and underdeveloped national roads significantly contribute to low market integration of the Kazakh wheat market with international markets. Besides a deficient grain transport infrastructure, large distances between grain producing and consuming regions strongly influence the degree of market integration of different regions in Russia. Infrastructural problems are also one of the main factors influencing market integration of Caucasian countries (i.e., Armenia, Azerbaijan, and Georgia) with the EU and international markets.

Fourth, strengthening trade relations through bilateral or multilateral trade agreements significantly contributes to market integration of the CIS countries. The establishment of the Eurasian Customs Union (EACU) facilitates the trade process between Belarus, Russia, and Kazakhstan. The results indicate that the EACU members have almost identical regional and international levels of integration. This is especially the case for pork markets. In addition, the results indicate strong market integration of CIS pork, beef, poultry, and milk markets with the EU. The main reason for almost full transmission of price shocks from the EU markets might be associated to numerous bilateral trade agreements between the EU and almost all CIS countries.

6. CONCISE SUMMARY

In this chapter, we have examined the extent to which certain CIS markets are integrated into regional and global agricultural markets. We analyzed the wheat, meat (pork, beef, and poultry), and milk markets using a distinctive data set we compiled from several sources for the CIS countries. The chosen markets are indicative of the primary agricultural industries in the CIS countries. Efficient agricultural markets are crucial for food security in the CIS countries and are therefore highly important for policymakers.

The price transmission research shows that regional integration within the CIS is most pronounced for pig and beef, with poultry and whole milk powder following closely behind. CIS markets are highly integrated into global agricultural markets for wheat and beef, but less so for pork and poultry. Overall, beef markets in the CIS countries are the most robustly interconnected within the region, as well as with the EU and the global market. Our findings suggest that domestic market support, trade policies, physical trade flows, infrastructure, and trade agreements are crucial factors influencing the market integration of CIS countries. Our findings suggest that markets with significant state support tend to have limited integration with international markets, or perhaps show a lack of integration, in contrast to markets without market support measures. Secondly, our findings suggest that CIS markets may be closely linked to both regional and international markets. Regional integration appears to be more influenced by physical trade flows (import dependency), whereas integration with international markets is more focused on establishing benchmark prices for products traded at the regional level. Furthermore, the inadequate infrastructure in most CIS nations greatly hinders market integration. Fourthly, enhancing trade links via bilateral or multilateral trade agreements greatly aids in the market integration of the CIS nations. Particularly true for members of the Eurasian Customs Union and CIS agreements with the EU.

7. AT A GLANCE

Utilizing a price transmission approach, we focus on price relationships between the countries of the Commonwealth of Independent States (CIS) and how price changes on the world agricultural market are transmitted to the domestic CIS markets. In this study, we establish a unique price data set on five different agricultural products (i.e. wheat, pork, beef, poultry and whole milk powder) observed in eight different CIS countries. The results of the price transmission analysis indicate that regional integration within the CIS is strongest for pork and beef, followed by poultry and whole milk powder. The integration of CIS markets in world agricultural markets is strongest for wheat and beef, whereas it is relatively low for pork and poultry. Furthermore, beef markets in the CIS countries are the strongest integrated within the region, with the EU and the world market. Overall, our results indicate that domestic market support and trade policies, physical trade flows between countries, infrastructure, and bilateral or multilateral trade agreements play a key role in market integration of thc CIS countries. These determinants should be particularly considered when designing recommendations for improving agri-food supply chain competitiveness in the CIS countries.

Appendix. List and Classification of Countries

	Domestic markets	World regional markets	World reference markets	
Countries	CIS	CIS	EU	World
Armenia	x			
Azerbaijan	x			
Georgia[a]	x			
Belarus	x			
Moldova	x			
Kazakhstan	x	x		
Russia x		x		
Ukraine[a]	x	x		
France			x	
Denmark			x	
Germany			x	
Netherlands			x	
Spain			x	
Oceania				x
Brazil				x
Australia				x
Argentina				x
Uruguay				x
USA				X

[a] Although Georgia and Ukraine are no longer CIS members, in this study, we still refer them as CIS countries due to their tight regional trade connections with other CIS members.

REFERENCES

1. European Commission. Analysis of price transmission along the food supply chain in the EU. Accompanying document to the communication from the Commission to the European Parliament, the Council, the European economic and social committee and the committee of the regions. Commission of the European communities, Commission stuff working document SEC 1450, Brussels, 28.10.2009http://groupedebruges.eu/sites/default/files/publications/downloads/publication16067_price_transmission_ec.pdf
2. Fackler P, Goodwin B. Spatial price analysis. In: Gardner B, Rausser G, editors. Handbook of Agricultural Economics. Vol. 1B. Amsterdam: Elsevier; pp. 971-1024. DOI: 10.1016/s1574-0072(01)10025-3.ch17
3. Tomek W, Robinson K. Agricultural Product Prices. 4th ed. Cornell Universtiy Press, Ithaca, New York, United States of America; 2003. ISBN-13: 978-0-8014-5230-7
4. Bluashvili A, Safaryan S. Global food price shocks transmission to local markets and welfare implications for Georgian households. Policy Paper 2014. The research is conducted in the framework of the project "Policy Research for Sustainable Growth", implemented by Konrad Adenauer Foundation in cooperation with PMC Research Center, Washington, United States of America. 2014
5. Katsia I, Mamardashvili P. Price transmission on wheat flour market in Georgia. Paper presented at the Samarkand Conference "Regional and International Cooperation in Central Asia and South Caucasus: Recent Developments in Agricultural Trade", Samarkand, Uzbekistan, 2-4 November 2016
6. Dickey DA, Fuller WA. Distribution of the estimators for autoregressive time series with a unit root. Journal of American Statistical Association 1979;74(366):427-431. DOI: 10.2307/2286348
7. Johansen S. Statistical analysis of cointegration vectors. Journal of Economic Dynamics and Control. 1988;12(2-3):231-254. DOI: 10.1016/0165-1889(88)90041-3
8. Stock JH. Asymptotic properties of least squares estimators of cointegrating vectors. Econometrics. 1987;56:1035-1056. DOI: 10.2307/1911260
9. FAO. Red Meat, Agribusiness Handbook. Food and Agriculture Organization of the United Nations, Rome, Italy. 2009
10. Acosta A, Ihle R, Robles M. Spatial price transmission of soaring milk prices from global to domestic markets. Agribusiness: An International Journal 2014;30(1):64-73. DOI: 10.1002/agr.21358

11. Svanidze M, Götz L, Djuric I, Glauben T. South Caucasus's wheat import from Black Sea region: What are the lessons for Central Asia? Paper presented at the 155th EAAE-Seminar: "European Agriculture towards 2030 – Perspectives for further East-West Integration", Kiev, Ukraine, 19-21 September 2016
12. Syzdykov R, Aitmambet K, Dautov A. AGRICISTRADE Country Report: Kazakhstan. Analytical Centre of Economic Policy in Agricultural Sector, Kazakhstan. 2015. Available from: http://www.agricistrade.eu/document-library
13. Volk T, Erjavec E, Rac I, Rednak M. AGRICISTRADE – Deliverable 2.3: Agricultural Policy in the European Union's Eastern Neighbours. 2015. Available from: http://www.agricistrade.eu/document-library
14. United Nation International Trade Statistics Database (UN Comtrade). Available from: www.comtrade.un.org. 2016
15. Akhramovich V, Chubrik A, Shymanovich, G. AGRICISTRADE Country Report: Belarus. Belarus: Research Centre of the Institute for Privatization and Management; 2015. Available from: http://www.agricistrade.eu/document-library
16. Götz L, Djuric I, Glauben T. Wheat export restrictions in Kazakhstan, Russia, and Ukraine: Impact on prices along the wheat-to-bread supply chain. In: Schmitz A, Meyers WH, editors. Transition to Agricultural Economies: The Future of Kazakhstan, Russia and Ukraine. CABI (Commonwealth Agricultural Bureaux (CAB) International), Wallingford, Oxfordshire, United Kingdom; 2015. pp. 191-203. DOI: 10.1079/9781780645353.0000.ch19
17. APK-Inform. AgriNews, various issues. Available from: https://www.apk-inform.com/en. 2016
18. United States Department of Agriculture (USDA). Available from: http://apps.fas.usda.gov/psdonline/psdquery.aspx. 2016
19. Irish Food Board. Available from: www.bordbia.ie. 2016
20. Djuric I, Götz L. Export restrictions – Do consumers really benefit? The wheat-to-bread supply chain in Serbia. Food Policy. 2016;63:112-123. DOI: 10.1016/j.foodpol.2016.07.002
21. Götz L, Djuric I, Nivievskyi O. Regional price effects of extreme weather events and wheat export controls in Russia and Ukraine. Journal of Agricultural Economics 2016;67(3):741-763. DOI: 10.1111/1477-9552.12167

09 THE CHALLENGES FACED BY SMALLHOLDER FARMERS: A CONTRIBUTING FACTOR TO CONTEMPORARY AGRICULTURAL VALUE CHAINS IN SOUTH AFRICA

1. INTRODUCTION: AN OVERVIEW

IAASTD estimates that 90%[1] of farms worldwide are Smallholder Farms (SHF) with less than 2 hectares of land. The Food and Agricultural Organization states that 92.3% [2]of farmers are small farms, with 83% being less than 2 hectares in size. Smallholder farmers remain the primary source of food for the global population. South Africa (SA) has an average farm size of 1400 [3]hectares. 1.6 million residents, constituting 3% of the population, are engaged in agriculture. Out of the total, 162,000 are classified as official rural farms, with just 30,000 operating as commercial farmers who produce more than 80% of the food in South Africa[4].South Africa has a highly developed national structure in comparison to emerging countries, as only 0.3% of its population is engaged in formal rural farming[5]. Only 0.06% of South Africa's population are involved in commercial farming, which is far lower than its proportion of the global population compared to the number of global farmers. [6, 7]. Approximately 26% of its people still resides in conditions where they lack reliable access to an adequate quantity of affordable, nutritious food. Africa encounters challenges in establishing modern agricultural value chains, especially concerning Smallholder Farmers (SHF) in South Africa. Researching these value chains and analyzing the possibility of economic marginalization of smallholder farmers throughout Africa is crucial.Africa's SHFs mostly practice low external input and organic agriculture [8, 9]. They sit on small parcels of land with high genetic diversity and under such conditions Altieri et al. [10, 11] argue they have less environmental impact than high external input (HEI) agriculture, as practiced by commercial industrialized farms [12, 13, 14]. Soil water conservation practices by these farmers are important for increased yields with significant benefits showing particularly in heat and drought stressed areas and Marenya and Barrett [15] argue that integrated natural resource and soil fertility management have positive feedback on SHF household wealth. Organic farming is ideally suited to SHFs say Hine and Pretty [16] because it relies on naturally available fertility inputs, requires less operating costs, delivers more diversity, and is more resilient to plant stress [16].

These SHFs are less dependent on large multinational corporations for input supplies [11], allowing developing nations to support a less import dependent trading system; whose trade arrangements otherwise typically favor larger farming operations. Because SHFs produce food more organically they can regionally supply crops of higher nutritional value [17, 18, 19, 20, 21]. Ponisio [22] argues that organic crops need not be less productive and can still produce between 91 and 96% that of conventional farming [22]. Other research shows that it can also outperform synthetic fertilizers on the continent, increasing yield 2–3 times while remaining drought resilient, produce less CO2 and use 2–7 times less energy [22, 23, 24].While social challenges persevere, environmental degradation, disturbed ecosystems, loss of topsoil, modern human sicknesses, large CO2 emissions [13, 25, 26, 27, 28], all due to large scale industrialized conventional agricultural that form the foundation of our modern food system complex, add to the problems and raise the question of whether our food system sustains long-term food security or rather puts it at risk for the benefit of a few large system operators. This lends weight to the argument that SHFs are a good possible route to a socially and ecologically just and intensified agricultural systems, with the potential, as Wigging [29] argues, to produce more food per hectare than large farms.

Yet, as Baiphethi and Jacobs [30] contend, only 10% of food for South African households is sourced from subsistence production. The best way in enhancing access to food is through subsistence and SHFs' food production, and because that access is direct, it would also drive down food prices [30].South Africa's government is prioritizing providing aspiring farmers with access to land through controversial land claims and helping them transition into larger commercial enterprises[31]. This has arguably led to numerous undertakings that are usually considered failed. Aliber and Hall suggest that rather than relying on the foundation of Smallholder Farmers (SHFs) to support the development of emerging black commercial farmers through costly projects[32, 33], efforts should be directed towards utilizing the abundant subsistence farmers in regions with established smallholder farm clusters in South Africa. Investing in these areas for adaptation, diversification, employment, and improved food security would be more effective in addressing widespread unemployment and poverty. The initiative involves establishing improved market connections to supermarkets, decentralized agro-processing to assist small-scale farmers, encouraging land renting, and implementing a more participatory agricultural strategy.[32, 33] We contend that smallholder farmers are inherently entrepreneurial as they generate surplus produce which they sell on informal markets and roadsides for profit. The government can harness this entrepreneurial spirit to improve the low yield outputs of smallholder farmers.

1.1. Problem Statement

SHFs in South Africa struggle to survive and participate in food value chains, which currently maintain a flow of capital funds through a few large VCPs to a few selected and preferred large crop producers. Not being able to take part in these value chains means exclusion from the capital markets and a general struggle for economic survival, while rural areas remain with the stigma of low opportunity for young people. The system fails SHF on multiple levels, but mostly on access to education, land, technology, market, and financial services [34].

1.2. Research Objective and Methodology

In this manuscript we identify existing blockages SHFs face in participating in modern agricultural value chains in SA, as well as which institutions, policies, and VCPs are responsible for such blockages. What blockages do VCPs experience themselves through circumstances exacerbating access problems and how do feedback loops in existing value chains further entrench obstructions inhibiting a participatory framework for SHFs?

This study was undertaken as an ethnographic research exploring business cultures and morals using qualitative semi-structured interviews. The selected research participants (VCPs) were based on their involvement in the food value chain and their general size and importance they played; they were not from any particular commodity type, however, because we also interviewed silos and millers of maize, answers from these VCP often hinged around maize, also a main crop type in South Africa [35].

The interviews were then transcribed to attain primary qualitative data and for the coding and categorizing we used grounded theory as an inductive systematic methodology to analyze qualitative data and give it conceptual structure through categorization of general themes emerging from the data [36, 37, 38, 39].

2. EXISTING KNOWLEDGE ON CHALLENGES FACING SHF

2.1. Smallholder Farmer Challenges in Accessing Micro-Loans

In the view of Delgado [40] barriers arise primarily because markets fail to present solutions, such as micro-credit, to rural African populations. Kirsten and van Zyl [41] argues that credit availability, among others, is either imperfect or missing as accessible service to SHFs, while Ortmann and King [34] argue that high transactional costs for VCP is also due to language barriers; only 36% of SHFs spoke English in two regions in KwaZulu-Natal. Thirty six percent small-scale producers in South Africa, farming on less than 2 ha land, indicated that missing access to credit was the biggest hurdle why they would not be able to access water for irrigation [42].

Bain & Company [43] illustrate examples of other tried and tested models with a list of best practices that underlie the success factor for scalable operations supplying micro-lending that was developed by the Grameen Bank. Swilling [44] criticizes Grameen Bank replicas, saying that such systems either need a critical mass of over 2 million members to finance bureaucracies, or otherwise charge high interest rates. In contrast Blewitt [45] compares the Grameen Bank's principles of loan business to that of Green Entrepreneurship.

Naess et al. believe it is more the responsibility of the government to make farming profitable for SHFs, and ensure access to land, water, and other inputs such as seeds and "approcredit" (appropriate credit) are available [46]. They criticize micro-loans in that they are too small and maintain a micro-status, limiting operational growth and appropriate infrastructure investment. They do not enhance labor's dignity nor do they raise the farmer's negotiation power against traders, or the ability to invest into storage and packaging to add value, altogether being unable to lift the farmer out of poverty [46].

2.2. Smallholder Farmer Challenges to Accessing Insurance

Kirsten and van Zyl [41] argue that access to insurance is imperfect or non-existent for smallholders. Many smallholder households in Lesotho do not have any form of insurance [47]. Because insurance is hardly affordable in developing countries, only 1% of households in low income countries have catastrophe coverage [48].

Linnerooth-Bayer and Mechler also argue that with climate change, associated risks insurance for SHFs will become more important. To Challinor et al. [49], SHFs are expected to have a greater exposure to climate risk due to increased variability of crop productions and the absence of well-functioning crop insurance services. In order to increase SHFs' resilience, the establishment of alternative insurance schemes are required [49]. An example is JuhudiKilimo in Kenya that adds insurance to its livestock loan at 4% of animal value to protect both itself and the insured livestock against illness and death [43].

2.3. Smallholder Farmer Challenges in Accessing Markets

A reason why SHFs with sellable surplus crops stay trapped in poverty is the lack of access to markets [50]. A few national retail stores have risen from 10% market share around 1990, to 60% today [51] and dominate the formal food market in South Africa [52]. Large retailers from this "supermarket revolution" work with non-family, corporate agriculture to develop production systems that, via audits, could claim attributes of environmental sustainability and food safety [53].

Campbell criticizes that certification systems brought about by an "audit culture" that merely serves the interests of retailers and poses considerable hurdles for third world producers, for many of whom it is impossible to adhere to the compliance requirements [53]. While Qaim and Rao [54] argue that this, together with more vertical integration and stricter standards (developing nations' following the pattern of developed nations), can have far-reaching effects on rural development, Snider et al. [55] argue that certification system run through cooperatives in Costa Rica had little financial benefits for the SHF and has induced no widespread change.

Small producers not only face competition from larger and successively growing producers in their own countries, but also from other countries through increased imports [56, 57]. A large retailer in SA, Shoprite, procures 90% of its fresh fruits and vegetables from large-scale farmers, while Angola's Shoprite stores get 99% of their produce from South Africa. Pick n Pay also sources 70% of their produce in Southern Africa, from South Africa [58]. On the other hand SHF form the "structural backbone" of the rural economy [59], yet the pressure is high on SHFs to adjust to shifts in technology and changes in the market, as well as competition from imports, and if widespread exclusion is observed, SHFs will face difficult times [59].

This will spell a disaster of "highest magnitude" argues Magdoff [60], particularly if the "supermarket revolution" trend continues to drive out SHFs globally, a disaster not only for billions currently involved with small scale agriculture Magdoff [60] (p. 116), but also for an entire era of more expensive energy and climate change exacerbated by large industrial agriculture. Parker [61] says: They "... are, in the harsh terms of globalization, superfluous." Modern food systems place SHF on the edge of survival, while others see smallholder farming increasingly as an essential route out of poverty [62].

2.4. Smallholder Farmers Milling and Agro-Processing

Rural, low-income areas have small, informal traders to which SHFs sell. Supermarkets on the other hand are able to undercut informal traders who are exposed to the risk of being ultimately crowded out by supermarket stores, increasing the risk of SHF survival [5, 63]. However, Godfray et al. [64] argue that there is an opportunity for food security in poorer regions with improved technology for small-scale food storage in a network of small scale traders, millers and producers.

Mahlogedi and Thindisa [65] argue that agro-processing for SHF creates added value and has the ability to improve the livelihoods of SHF. However, they also say that would require sufficient human and social capital from the SHF. Local and subsistence food production is the best readily available route to food security in South Africa [30]. Baiphethi and Jacobs also believe that "rural households continue to value the pursuit of farming activities and that subsistence production is important to improve household food security" [30]. Similarly, Jayne et al. believe that any "realistic discussions of poverty alleviation strategies in Africa need to be in the context of access to land, [...] there is a strong relationship between access to land and household income" [66].

3. VALUE CHAIN RESEARCH FINDINGS

Our research findings from the interviewed VCP are summed up as follows:

3.1. Banks

We interviewed four of the largest commercial banks in South Africa, all of whom have, in one way or another, engaged with supplying credit to SHFs or emerging commercial farmers. Their views around what constituted micro-lending or SHF loans, varied drastically, ranging from loans of R500 for one bank to R100,000 for another bank. One bank said a R3m turnover was a minimum limit.

While one bank had a classic agricultural micro-loan, all other banks did not have a product tailored for SHFs. Two banks responded saying that SHF could use their bank's private loans of R500 instead of their classic agricultural loans which were designed for larger commercial farmers. One bank had an engagement another large VCP, through which SHF could access micro-loans for crop production.

The reasons why banks were unwilling to engage SHFs with products and services, they argued that their commercial mandate was focused on larger commercial farms that had collaterals, track records and economies of scale. Banks had an obligation, all respondents said, to find out where risk reducing factors existed that constituted repayment ability and affordability of the loan. This SHF with their experience, were not able to do. Access to decent financial histories to support an application for a loan were generally missing, without which it would be reckless lending and prohibited by the SA Credit Act.

Three of the four banks agreed with our question, whether crop in itself would suffice as an alternative form of collateral, but they said it would then need to be attached to an offtake agreements backed by an insurance that could pay the loan in case of failure.

The two banks that did not supply micro-loans said that access to common land would not qualify them for loans, it would have to be documented land tenure in form of ownership or lease agreements. Support in the form of a mentor, or a business development, most banks said would also constitute a de-risk for them in their decision making of supplying micro-loans. To them financial acumen mitigated production-, marketing-, and financial risks and would validate the supply of future micro-loan products, if education were able to drive a change in SHF mentality.

Historically, two banks had made bad experiences, where they had to write off debt: one bank lost 60% of a R3bn exposure. Subsequently one of the two banks backed off from supplying micro-loans and is not planning any new products tailored to SHFs. The other bank continues to supply SHFs with micro-loans. This bank also indicated that there were a host of challenges with micro-loans one being that their transaction costs were unprofitably expensive, making it a philanthropic offering to their bank.

All banks agreed that certain processes, compliance requirements and general customer costs they needed to bear made it unlucrative to serve very small loans. Getting FICA (Financial Intelligence Centre Act) requirements in place for clients far out in rural areas compounded the problem and with often missing identification documents, it would include having to establish identity and domicile in rural Africa, this alone disqualified many applicants. As opposed to one large farmer, with a R2m credit and little risk, 1000 SHFs with a credit each of R2,000, would be hugely more expensive, and at the same time expose them to high levels of risk and default because the loans were relatively unsecured.

One bank channeled their farmer micro-loan offering through their private loans division and not the agricultural business loans. They said their loans were used to buy seeds, repay debt, and to transport produce to market. First time applicants to their loans did not have it as easy as second or third time applicants. First time applicants would typically start with R500 and a tenure of 6 months, at prime plus 3%. In later rounds, the loans could increase to R1000 over 12 months and even up to R5000 with an established track record.

The three banks not supplying micro-loans argued that although they wanted to fund that market space, there was no specific model that would make business sense. The fourth bank that did supply micro-loans was offering it as a part of a total value offering and part of the financial services charter, assisting emerging agriculture. In Kenya, these banks stated, it had a successful micro-loan scheme because it had access to the applicants via cell-phone technology.

On our questions of government grants, one bank thought that knowledge and education were factors that made a farmer productive and would be better than grants. That bank argued that grants could kick-start businesses and create a success environment if applied correctly, but criticized the government for running unsuccessful grant projects, due to the disjuncture within government departments, that made SHF dependent on continued grant funding. Two of the four respondents were of the opinion that the primary responsibility for developing emerging agriculture lay with the state and the Land Bank. In this light the Kula-Scheme was also criticized for burdening their bank with bureaucracy in form of qualifying criteria and project management.

The principle opinion of two banks was that, South Africa needed an integrated policy framework, where the Land Bank, with an increased mandate for SHFs, and other organs of state, all played a part in driving agricultural-entrepreneurship to alleviate poverty. Corporate social investment (CSI) funding could assist, but currently is "wasted" money, as most of the projects ended up as write-offs.

One bank suggested that the government could also institute a guarantee fund for commercial banks to claim from, if bad debt and failures specific to this market occurred. Preferably, this should work in combination with a farm-level two-year grant for inputs and education. Most banks perceived existing extension offices to be unsuccessful, nonetheless, all banks believed that a form of mentorship for life-, financial-, and farming skill was essential to make sure money was utilized responsibly. The banks could not provide extension services, as they would then take default risk to the loan they supplied.

Only one bank showed a concern around our question of land ownership. They suggested the creation of a system that would enable SHF to have title deeds, which would develop a local property economy where successful SHFs would be able to buy additional 2 ha close by. Having title to their land would enable them to fully secure their facilities and grow their business. This would create a spirit of entrepreneurship among SHFs this bank argued. Another bank claimed that land ownership would help as a collateral, it was of less importance to them if other collaterals were in place, such as access to market and offtake, coupled possibly to insurance.

Although all four banks quoted FICA and the Credit Act as posing challenges to supply micro-loans to SHFs, only one bank suggested a change of that legislation. One bank argued that policy makers should rather look for successes in other African countries where cooperatives created successful farming ecosystem that enabled the successful supplying of micro-loans. A government guarantee scheme for drought failure, for example, to de-risk banks, could work through cooperatives and target not just one, but 20 or 30 farmers, with one collateral manager who helped control production schemes. Cooperatives worked in the mind of this bank, because of scale, where many SHFs pooled their maize together, reducing transactional costs to market. The Land Bank could assist they said.

A cooperative, another bank believed, can be very important, if it is a commercial co-op with a good existing integration into the value chain that can function as a service provider in terms of information flow, technical expertise, and possibly a funds disseminator. This bank though, also said that it was not necessary to pool every SHF into cooperatives, especially if the SHFs had access to good local market they could serve.

3.2. Supermarkets

Our interview base covered three major retailers in South Africa, all of whom had some form of engagement with SHFs or emerging farmers. The authors planned and conducted the interview with senior employees of these organizations responsible for the purchasing of farm produce.

One of the retailers said that they almost exclusively bought commercial volumes and that SHFs were ever only going to be a very small part of that supply. They supplied reliable commercial growers with growing programs to which both parties committed. A SHF in comparison was an unreliable source for multiple reasons on which they could not rely. This retailer also said that SHFs lend themselves to niche, high value, out of season production and could make a success there. The new BEE code, requiring them to source more from SHF, would change things dramatically, but would nonetheless pose a challenge to them.

The second retailer said that they worked with fresh produce SHFs, mainly in Limpopo, through their formal pack-houses and central procurement. This retailer also said that they allowed their store owners to procure, outside of their central procurement system, directly from smaller farmers. This was mostly done without cooperatives being present, but nevertheless resulted in problems with consistency and uncoordinated supply, which the store owners accepted because of the higher margins they made by procuring directly from SHF.

That retailer considered SHF farmers as ones with more than 5 ha of land, and they did not think one could farm sustainably on 2 ha of land or less, other than maybe through a cooperative system. For these SHF the food safety and quality requirements like Global GAP, Tesco Nature's Choice, or GFSI, are almost impossible to adhere to they said and therefore they created their own "Local-GAP" for SHF as compliance capacity-building with a chance to step up to Global GAP. All three retailers are concerned about SHFs ability to comply with food safety standards, which was essential and needed to stay in place.

The third retailer claimed that they did not work with SHFs as their scale was too small and that they would need to pool 50 or 60 farmers together and manage them to get the produce they needed. They have had no SHFs projects in the past and are not planning any in the future. While this was their cooperate approach they said, their individual franchises would be able to procure directly from SHFs in their vicinity, which even then in their opinion would be too small in scale. In response to how government could get involved, one retailer said that government should facilitate systems where successful commercial farmers mentored SHFs collaboratively alongside a market access to retailers. To them, the retailers have the expertise, the network and accountability, while the government has the money to facilitate such engagements.

The second retailer was of the opinion that there were three levels with which the government should engage. First, to assist SHF in attaining finance, second, to raise the skill of SHFs to run better farms and businesses, and third, to assist with entry level food safety and compliance schemes.

The third retailer was of the opinion that money was not needed, and that it was the infrastructure and system around SHFs that role players, such as banks, retailers, and especially government, needed to create. This retailer said that it would need to be a whole number of things that are required to fall in place, and that somebody needs to take control of and manage it; best would be the government.

3.3. Traders with Silos and Milling

Both interviewed traders were not buying from SHFs. They said they were mostly procuring from other traders and only in a very few cases directly from usually larger farmers. The new Black Economic Empowerment Code (BEE) in SA also did not require them to buy from SHFs. While both traders said that they were not giving credits, seeds, or input supplies to SHFs, one trader said that such support might be possible through their Enterprise Development funding, but that there was additionally also a new department in the making that was going to specifically focus on socially just procurement in the future for their company.

One of the traders said they would love to support small-scale farmers and pay them market-related prices, but these would need to supply trucks loads of greater than 30 tons for effective scale. For SHFs, getting to the market would be the biggest challenge, they said; they also did not think that any other traders wanted to take any risk with SHFs.

On the question of how government could play a role, one trader said it could assist in pulling together many small-scale farmers into a corporative, where it became viable to have one contract with a community to buy 30 or 100 tons, where minimum truckloads were 35 tons to get the crop to Randfontein. The second trader said the government needed to empower SHFs first, with subsidies to decrease input costs and secondly with field extension to increase outputs. Other than that, this trader responded, the government should just stay out of economics.

3.4. Insurers

We interviewed three of the largest insurers in SA, who together covered more than 80% of the insurance market in South Africa. Two of the three insurers claimed that they already had a product with which they served SHF, but with a focus on livestock and not crop insurance. The third insurance company said that they currently had no micro-insurance product for SHF, but that they have had engagements in the past. To this insurer the traditional underwriting model suited commercial farmers and not SHF because historic data and proof of affordability on their balance sheets was missing.

Nevertheless, this insurer indicated that they were busy with the Land Bank and the International Climate Insurance Fund to build a new product for SHF. Another insurer indicated that it was busy with a National Emergent Red Meat Producers' Organization (NERPO) and Grain SA project. Generally, all three insurances agreed that there was not sufficient historic financial data from SHF that would enable them to supply classical crop insurance. One insurer said that the high capital backup requirement of 120% was costly to run even for commercial farmers. The assessment of doing pre-emergence, post-emergence and loss-reporting further drove up the costs, in particular for SHF. Subsequently two insurers were of the opinion that the Financial Services Board (FSB) should deregulate indexed insurance which would vastly reduce costs of supplying insurance to both commercial and SHF, in which case simple climatic models would trigger pay outs. However, this would still need to be tested and two insurers indicated that attaining meteorological data for rural SHF was in most cases very difficult.

Two insurers also indicated that South Africa was one of the very few countries in the world where the government did not subsidize agricultural insurance. One insurance said that the government should think about subsidizing insurance as an alternative to a national backup fund, as such a subsidy would be able to reduce the cost burden of the high backup capital to be carried by the insurers. The private industry backup capital could replace the idea of having a national backup fund and with that reduce costs of operating premiums. The fall-outs would be easier to carry due to a potentially much larger client base, where even SHF could be served easier.

Other mentioned challenges were the transacting of payments from and to SHF and the expectations both parties had on and against a claim; unlike commercial farmers who understood insurance concepts very well. Subsequently a cost-effective service delivery is a challenge, particularly on products of less than R100, where agents would make next to no money.

4. DISCUSSION

4.1. Summary of Data

The collected qualitative data was transcribed and systematically coded and categorized using grounded theory methods to gain structure of more quantitative nature, summed up here into two tables (Table 1, Table 2):

Table 1.Serving SHFs today and tomorrow?

	Serving SHFs today?		Serving SHFs in future?	
SHFs	Yes (%)	No (%)	Yes (%)	No (%)
Banks [4]	25.0	75.0	50.0	50.0
Insurers [3]	66.7	33.3	100.0	0.0
Traders [2]	0.0	100.0	50.0	50.0
Retailers [3]	0.0	100.0	66.7	33.3
TOTAL	22.9	77.1	66.7	33.3

Table 2.Limitations inhibiting both SHFs and VCPs.

	Limitations experienced by both SHFs and VCPs							
SHFs	Funding (%)	Land ownership (%)	Education (%)	Logistics (%)	Compliance (%)	FICA, credit act & FSB (%)	Access to market (%)	Cooperative leadership (%)
Banks [4]	87.5	50.0	100.0	90.0	95.0	50.0	88.3	90.0
Insurers [3]	80.0	33.3	86.7	80.0	53.3	66.7	71.1	80.0
Traders [2]	90.0	0.0	90.0	70.0	60.0	n.a.	80.0	100.0
Retailers [3]	66.7	33.3	73.3	100.0	93.3	n.a.	88.9	90.0
TOTAL	81.0	29.2	87.5	85.0	75.4	58.3	82.1	90.0
In agreement (stdev)	10.5	21.0	11.0	12.9	21.8	11.8	8.4	8.2

While 77.1% of surveyed VCP do not engage with SHFs today, the willingness to engage in future with SHFs is three times the current engagement. This willingness was limited to the assumption that other VCPs would also start to engage SHF more than they did now. In other words a collective effort would entice a joint effort supported by 2/3rd of all VCPs.

Of all limitations landownership is the last issue why VCP do not to engage with SHF. However, relative to other limitations a higher standard deviation indicates that, in particular Banks disagree, which is understandable, as they use land ownership as collateral. Second least important to the financial institutions were FICA, credit act and FSB regulations. More important are compliance with food safety and Gap standards; both banks and retailers vote them as second most important. All interviewees agree that cooperative leadership is the most important limitation that SHF and their own institutions face. Ma and Abdulai [67] have also shown in their studies that cooperative membership has a significant positive impact for SHF on yields, net returns and household incomes.

Next important to all interviewees are education and logistics and after that access to market and funding where for both the VCPs also seem to agree, with a small standard deviation.

The reader must keep in mind that due to the concentration of large organizations in the South African value chain the sample size is relatively small. However, the comparatively low standard deviation lends value to the research findings; in particular, where more than 80% of VCPs agree that cooperative leadership, access to market, education and funding are the most important limitations that need solutions.

4.2. Policy Debate

While some VCP had "spectacular" failures working with SHFs in Africa, most do not engage SHFs to avoid risk. However, the willingness to engage SHFs in future within a more functional economic system favoring SHFs is high.

From our categorized limitations, all limitations increased the risk for any of the VCP to engage with SHFs. Subsequently reducing the exposure to risk for VCPs is likely to create a more interesting environment for them to working with SHFs. Policy therefore should focus on risk reducing concepts and limitations as prioritized in Table 2.

Within the current legal framework of the Credit Act, the rulings for insurance pay-outs by the Financial Services Board (FSB), and FICA, the financial industries, banks and insurances are limited to serve larger commercial farms and forced to ignore SHFs. Yet more important to the financial services as well as the traders and retailers were limitations related to cooperative leadership, access to market, education, funding, and also logistics. Compliance to food safety standards was a very important limitation to banks and retailers. All these limitations were more important to the VCPs than legislative and regulative requirements or land ownership.

This raises the question, whether the SA, and other African, governments' policies are on an effective road by largely focusing on land ownership, often politically contentious. To all VCPs land ownership was by far the smallest concern. Even to banks, to whom an obvious de-risk factor is land as a collateral, land tenure with the ability to create profit was more important to them than land ownership, which neither is a guarantee for good land custodianship nor profitability. Lease agreements from communal land in a more traditional environment would be sufficient to the banks. The concept of dropping land ownership policies in favor of communal lease tenure and cooperative engagement with commercial farmers pose a challenge to many policy makers in SA, as most commercial farmers are considered historic rivals. Blignaut et al. [68] found in their field report, that only 1.8% of their respondent thought that landownership for emerging farmers was an important part for policy considerations.

Understanding the challenges of the system from all conducted interviews, for government in particular, we can say policy should focus on cooperative leadership in combination with larger commercial farmers and off-takers to solve limitations in form of access to market and logistics. Making funding available, for such new systems, that did not create SHFs dependencies on grants, together with education would have the potential to solve the most important problems facing SHFs. If education were then additionally to focused on low external inputs and agroecological principles, reducing the need for expensive inputs, it would not only reduce the dependency on multinational corporations, their product imports and complicated logistics, it would increase local food sovereignty and reduce the risk of engagement for financial institutions due to less credit needs.

Such a systems approach would likely raise the interest of existing VCPs to engage with SHFs, because as risk reduces, a chance for profitability increases. Two-third of all VCPs have indicated that they would increase their engagement within months after they saw risk was reduced and other VCPs started working with SHFs. However, any such commercialization effort to Poole et al. [69] should have a decided mindset and must consider local complexities in order to get into the "hearts and minds" of the SHF, as otherwise it may not be an attractive profession to pursue or make a success of.

5. CONCISE SUMMARY

We have questioned whether South Africa's current food system can maintain long-term food security, given the current food structure influences economies and supports huge industrialized agriculture, while neglecting small and micro-food producers. We also contend that other African nations, in their efforts to modernize food value chains, should not overlook the opportunity for smallholder farmers (SHFs) to profit, as they possess significant food security potential. Our interview findings reveal that many systemic challenges are causing widespread opposition from multiple industries, especially across the SA value chain, to collaborate with SHFs. Lacking access to various components of value chains, such as input items, micro-loans, micro-insurance, education, and markets, leads to exclusion from revenue opportunities and creates economic challenges for survival. However, as stated in our research study, South Africa's SHF possesses an inherent entrepreneurial spirit that presents significant untapped potential.

Risk-averse venture capital partners avoid investing in startup high-failure industries due to their own experiences and the perceived high risk involved. However, the majority of VCPs are still extremely interested in enhancing their participation with SHF in the future, provided that a new system decreases the risk. We have demonstrated how the government could mitigate the primary risks and constraints, ranked in order of significance as follows: cooperative leadership (90%), education (88%), logistics (85%), market access (82%), funding (81%), and food safety compliance (75%). These aspects were deemed more crucial by interviewees compared to legislation, regulatory mandates, and land ownership. The government should prioritize economic viability over land ownership when promoting proper land custodianship and profitability. Funding should be directed towards establishing cooperative leadership with commercial farmers to provide access to markets, logistics, and education on low external input farming methods. These methods can reduce risk and reliance on credit and imported inputs, while enhancing food resilience in rural areas and the economic viability of smallholder farmers.

6. AT A GLANCE

The long-term food security of smallholder farmers is at risk due to the increasing dominance of modern food value chains, especially in Africa. SHF should play a role in driving support for a greener, lower carbon economy that generates jobs and enhances human well-being within a sustainable and socially inclusive economic development framework.

REFERENCES

1. IAASTD. Agriculture at a Crossroads: Sub-Saharan Africa (SSA) Report. Vol. V. Washington DC; 2009
2. FAOSTAT. Global distribution of agricultural land [Internet]. FAO of the UN. 2014. Available from: http://faostat3.fao.org/faostat-gateway/go/to/home/E [Cited: April 20, 2005]. pp. 1-10

3. Ramaila M, Mahlangu S, du Toit D. Agricultural Productivity in South Africa: Literature Review. 2011
4. May J, Carter M. Agriculture: Analysis of the NIDS Wave 1 Dataset Discussion Paper No. 6. 2009
5. Heijden V, Vink N. Good for whom? Supermarkets and small farmers in South Africa—A critical review of current approaches to increasing access to. Agrekon. 2013;52(1):68-86
6. Labadarios D, ZJ-R MH, Steyn NP, Gericke G, EMW M, Davids YD, et al. Food security in South Africa: a review of national surveys. Bulletin of the World Health Organization [Internet]. 2011 Dec 1 [Cited 2014 Nov 9];89(12):891-899. Available from: http://www.pubmedcentral.nih.gov/articlerender.fcgi?artid=3260897&tool=pmcentrez&rendertype=abstract
7. SANHANES-1. The South African National Health and Nutrition Examination Survey. Cape Town: HSRC Press; 2013
8. De Schutter O. Report Submitted by the Special Rapporteur on the right to food. Development. 2010, December. p. 21
9. IFAD-UNEP. Smallholder, food security, and the environment. International Fund for Agricultural Development and United Nations Environments Programme; 2013
10. Altieri M. Agroecology, small farm and food sovereignty. Monthly Review. 2009;61(3):102-113
11. Altieri M, Funes-monzote F, Petersen P. Agroecologically efficient agricultural systems for smallholder farmers: Contributions to food sovereignty. Agronomy for Sustainable Development [Internet]. 2011 Dec 14 [Cited 2014 Apr 28];32(1):1-13. DOI: . Available from: http://link.springer.com/10.1007/s13593-011-0065-6
12. Altieri M, Nicholls CI. Chapter 7: Enhancing the productivity of Latin America. Traditional peasant farming. In: Agroecology and the Search for a Truly Sustainable Agriculture. 2005
13. Fischer J, Brosi B, Daily GC, Ehrlich PR, Goldman R, Goldstein J, et al. Should agricultural policies encourage land sparing or wildlife-friendly farming? Frontiers in Ecology and the Environment [Internet]. 2008 Sep [Cited 2014 May 5];6(7):380-385. Available from: http://www.esajournals.org/doi/abs/10.1890/070019
14. Diaz RJ, Rosenberg R. Spreading dead zones and consequences for marine ecosystems. Science [Internet]. 2008 Aug 15 [Cited 2014 Apr 30];321(5891):926-929. Available from: http://www.ncbi.nlm.nih.gov/pubmed/18703733

15. Marenya PP, Barrett CB. Soil quality and fertilizer use rates among smallholder farmers in western Kenya. Agricultural Economics. 2009;40(5):561-572

16. UNEP-UNCTAD. Organic Agriculture and Food Security in Africa [Internet]. Geneva & New York. 2007. Available from: unctad.org/en/docs/ditcted200715_en.pdf [Cited: March 1, 2014]

17. Chassy AW, Bui L, Renaud ENC, Van Horn M, Mitchell AE. Three-year comparison of the content of antioxidant microconstituents and several quality characteristics in organic vs conv. Journal of Agricultural and Food Chemistry [Internet]. 2006 Oct 18 [Cited 2014 May 2];54(21):8244-8252. Available from: http://www.ncbi.nlm.nih.gov/pubmed/17032035

18. Res M, Dangour AD, Allen E, Lock K, Uauy R. Nutritional composition & health benefits of organic foods—Using systematic reviews to question the available evidence. The Indian Journal of Medical Research. 2010 Apr;131(April):478-480

19. Huber M, Rembiałkowska E, Średnicka D, Bügel S, van de Vijver LPL. Organic food and impact on human health: Assessing the status quo and prospects of research. NJAS - Wageningen Journal of Life Sciences [Internet]. 2011 Dec [Cited 2014 May 2];58(3–4):103-109. Available from: http://www.sciencedirect.com/science/article/pii/S1573521411000054

20. Rodale Institute. Nutritional Value of Fruits, Veggies is Dwindling—Health—Diet and nutrition—msnbc.com [Internet]. Rodale Institute. 2012. Available from: http://www.nbcnews.com/id/37396355/ns/health-diet_and_nutrition/t/nutritional-value-fruits-veggies-dwindling/#.U4SdkXKSySp [Cited: May 27, 2014]

21. Volakakis N, Seal C, Sanderson R, Dominika S, Stewart GB, Benbrook C, et al. Higher antioxidant and lower cadmium concentrations and lower incidence of pesticide residues in organically grown crops: a systematic literature review and meta-analyses. British Journal of Nutrition [Internet]. 2014 Jun 26 [Cited 2014 Jul 11]:1-18. Available from: http://www.ncbi.nlm.nih.gov/pubmed/24968103

22. Ponisio LC, M'Gonigle LK, Mace KC, Palomino J, de Valpine P, Kremen C. Diversification practices reduce organic to conventional yield gap. Proceedings of the Royal Society B: Biological Sciences. 2014 Dec 10;282(1799):20141396

23. Edwards S. Greening Ethiopia series. Sci Soc. 2004;23:4-8

24. Rodale Institute. Regenerative Organic Agriculture and Climate Change: A Down-to-Earth solution to Global Warming. Vol. 1400. Kutztown; 2014

25. Adhya T, Sharma P, Gogoi KA. Mitigating greenhouse gas emission from agriculture. In: Climate Change and Crops. Berlin, Heidelberg: Springer-Verlag; 2009. pp. 329-344

26. Ewers RM, Didham RK. Pervasive impact of large-scale edge effects on a beetle community. Proceedings of the National Academy of Sciences. 2008;105(14):4-7
27. Pimentel D, Harvey C, Resosudarmo P, Sinclair K, Kurz D, McNair M, et al. Environmental and economic costs of soil erosion and conservation benefits. Science. 1995;267(5201):1117-1123
28. Pretty J. Agricultural sustainability: Concepts, principles and evidence. Philosophical Transactions of the Royal Society of London [Internet]. 2008 Mar 12 [Cited 2014 Apr 30];363(1491):447-465. Available from: http://www.pubmedcentral.nih.gov/articlerender.fcgi?artid=2610163&tool=pmcentrez&rendertype=abstract
29. Wiggins S. Can the smallholder model deliver poverty reduction and food security for a rapidly growing population in Africa? Brighton; 2009. Report No: 8
30. Baiphethi MN, Jacobs PT. The contribution of subsistence farming to food security in South Africa. Agrekon. 2009 Dec;48(4):459-482
31. Greenberg S. The Disjunctures of Land and Agricultural Reform in South Africa: Implications for the Agri-Food System2013, (August). pp. 1-29
32. Aliber M, Hall R. Programme to support pro-poor policy development (PSPPD). In: Development of Evidence-Based Policy Around Small-Scale Farming. 2010
33. Aliber M, Hall R. Support for smallholder farmers in South Africa: Challenges of scale and strategy. Development Southern Africa. 2012;29(4):548-562
34. Ortmann GF, King RP. Agricultural cooperatives II: Can they facilitate access of small-scale farmers in South Africa to input and product markets? Agrekon. 2007;46(2(March 2015)):219-244
35. DAFF. Crops and Markets. Vol. 892008
36. Pettigrew SF. Ethnography and Grounded Theory: A Happy Marriage? http://www.acrwebsite.org [Internet]. 2016. Available from: http://www.acrwebsite.org/search/view-conference-proceedings.aspx?Id=8400 [Cited: August 21, 2016]
37. Savin-Baden M, Major CH. Qualitative Research: The Essential Guide to Theory and Practice [Internet]. London and New York: Routledge; 2012 [Cited 2016 Aug 21]. 569p. Available from: http://www.routledge.com/books/details/9780415674782/
38. Gibbs. Core Elements. Part 2: Grounded Theory [Internet]. 2016. Available from: https://www.youtube.com/watch?v=dbntk_xeLHA&feature=related [Cited: August 21, 2016]

39. Glaser BG. Grounded Theory Institute—The Grounded Theory Methodology of Barney G. Glaser, PhD—What is GT? [Internet]. 2014. Available from: http://www.groundedtheory.com/what-is-gt.aspx [Cited: August 21, 2014]
40. Delgado CL. Sources of growth in smallholder agriculture integration of smallholders with processors in sub-Saharan Africa: The role of vertical and marketers of high value-added items. Agrekon. 1999;38(sup001):165-189
41. Kirsten JF, Van Zyl J. Defining small-scale farmers in the South African context. 1998;37(4):551-562
42. Bryan E, Deressa TT, Gbetibouo GA, Ringler C. Adaptation to climate change in Ethiopia and South Africa: Options and constraints. Environmental Science & Policy. 2009;12(4):413-426
43. Bain & Company. Growing Prosperity: Developing Repeatable Models to Scale. 2014
44. Swilling M. Hear the Forest Grow: SDI Evaluation—Africa. Stellenbosch; 2005
45. Blewitt J. Understanding Sustainable Development. London: Earthscan; 2008
46. Naess A, Coulibaly A, Alhaji J, Diop A, Solheim E, Webber F, et al. . In: Naerstad A, editor. Africa can Feed Itself. Oslo: The Development Fund Norway; 2007
47. Ziervogel G. Targeting seasonal climate forecasts for integration into household level decisions: The case of smallholder farmers in Lesotho. The Geographical Journal. 2004;170(1):6-21
48. Linnerooth-Bayer J, Mechler R. Insurance for assisting adaptation to climate change in developing countries: A proposed strategy. Climate Policy. 2007;6(2006):1-17
49. Challinor A, Cochrane K, Howden M, Iqbal MM, Wgii I, Chapter AR, et al. IPCC WGII AR5. 2014. Chapter 7
50. Magingxa LL, Alemu ZG, van Schalkwyk HD. Factors influencing access to produce markets for smallholder irrigators in South Africa. Development Southern Africa [Internet]. 2009;26(1):47-58. Available from: http://www.tandfonline.com/doi/abs/10.1080/03768350802640081
51. Andersson CIM, Kiria CG, Qaim M, Rao EJO. Following up on smallholder farmers and supermarkets. Glob Food Discuss Pap. 2013;23:1-29
52. DAFF. Maize Market Value Chain Profile. Arcadia; 2012
53. Campbell H. The rise and rise of EurepGAP: European (re)invention of colonial food relations? International Journal of Sociology of Agriculture and Food. 2005;13(December):1-19

54. Qaim M, Rao EJO. The supermarket revolution and smallholder farmers. The International Journal for Rural Development. 2012 Apr;2011

55. Snider A, Snider A, Kraus E, Sibelet N, Bosselmann AS. Influence of voluntary coffee certifications on cooperatives' advisory services and agricultural practices of smallholder farm. The Journal of Agricultural Education and Extension [Internet]. 2017:1-19. Available from: http://dx.doi.org/10.1080/1389224X.2016.1227418

56. Louw A, Vermeulen H, Kirsten J, Madevu H. Securing small farmer participation in supermarket supply chains in South Africa. Development Southern Africa [Internet]. 2007 Oct;24 (4):539-551. Available from: http://www.informaworld.com/

57. openurl?genre=article&doi=10.1080/03768350701577657&magic=crossref%7C%7CD 404A21C5BB053405B1A640AFFD44AE3

58. Muchopa CL. Agricultural value chains and smallholder producer relations in the context of supermarket chain proliferation in Southern Africa. International Journal of Value Supply Chain Management [Internet]. 2013 Sep 30 [Cited 2015 Jan 7];4(3):33-44. Available from: http://www.airccse.org/journal/mvsc/papers/4313ijmvsc04.pdf

59. Crush J, Frayne B. Supermarket expansion and the informal food economy in Southern African cities: Implications for urban food security. Journal of Southern African Studies [Internet]. 2011 Dec [Cited 2015 Jan 8];37(4):781-807. Available from: http://www.tandfonline.com/doi/abs/10.1080/03057070.2011.617532

60. McCullough EB, Pingali PL, Stamoulis KG. The Transformation of Agri-Food Systems: Globalization, Supply Chains and Smallholder Farmers. Food & Agriculture Org.; 2008

61. Magdoff F et al. Renewable Agriculture and Food Systems [Internet]. 2007 Jul 4 [Cited 2014 Feb 7];22(2):109. Available from: http://www.journals.cambridge.org/abstract_S1742170507001846

62. Parker G. The Megacity: Decoding the Legacy of Lagos. The New Yorker [Internet]. 2006. Available from: www.soystats.com/2005/Default-frames.htm

63. Page S, Slate R. Small producer participation in global food systems: Policy opportunities and constraints. Development Policy Review. 2003;21(5–6):641-654

64. van der Heijden T. Good for Who? Supermarkets and Small Farmers in South Africa—A Critical Review of Current Approaches to Market Access for Small Farmers in Developing Countries. Stellenbosch University; 2010

65. Godfray HCJ, Beddington JR, Crute IR, Haddad L, Lawrence D, Muir JF, et al. Food security: the challenge of feeding 9 billion people. Science [Internet]. 2010 Feb 12 [Cited 2014 Jul 10];327(5967):812-818. Available from: http://www.ncbi.nlm.nih.gov/pubmed/20110467

66. Mahlogedi L, Thindisa V. Participation by Smallholder Farming Entrepreneurs in Agro-Processing Activities in South Africa. Johannesburg; 2014

67. Jayne TS, Yamano T, Weber MT, Tschirley D, Benfica R, Chapoto A, et al. Smallholder income and land distribution in Africa: Implications for poverty reduction strategies. Food Policy [Internet]. 2003 Jun [Cited 2014 May 10];28(3):253-275. Available from: http://linkinghub.elsevier.com/retrieve/pii/S0306919203000460

68. Ma W, Abdulai A. Does cooperative membership improve household welfare? Evidence from apple farmers in China. Food Policy [Internet]. 2016 Jan 1 [Cited 2018 Jan 21];58:94-102. Available from: http://www.sciencedirect.com/science/article/pii/S0306919215001396

69. Blignaut JN, De Wit MP, Midgley S, Crookes DC, Knot J, Drimie S, et al. Sustainable Farming as a Viable Option for Enhanced Food and Nutritional Security and a Sustainable Productive Resource Base—Deliverable 4 Field Report. Pretoria; 2015

70. Poole ND, Chitundu M, Msoni R. Commercialisation: A meta-approach for agricultural development among smallholder farmers in Africa? Food Policy. 2013;41:155-165

10 SMALL FARMERS' INTEGRATION INTO VALUE CHAINS IN EASTERN EUROPE AND CENTRAL ASIA

1. INTRODUCTION: AN OVERVIEW

In Eastern Europe and Central Asia (EECA), the whole food chain, including farm suppliers and retailers, experienced a significant disruption in commercial connections at the start of the transition process. Supply disruptions and substandard food products are now routine occurrences. Simultaneously, shifts in consumer preferences and the influx of foreign companies, particularly retailers and processors, required substantial reforms and modifications in the food commodity chains' framework to address these issues. Policymakers have to exert significant effort to organize food commodity chains and ensure improved quality and safety in the food supply. Typically, when changes occur, there is a need for extensive institutional modifications. As a result, a higher level of privately initiated vertical coordination among transaction partners in the supply chain has become a common method to address issues related to inadequate supply and low quality[1]. The new procurement processes require suppliers to ensure uninterrupted product flows and provide products of a specific quality. Therefore, local manufacturers must meet both the quantity and quality standards.

Various studies indicate that retailers, processors, and governments favor the expansion of farm size for various reasons [2, 3, 4, 5]. Despite this, merchants and processors must work together with small farmers because these farmers are crucial for meeting the necessary amounts in certain transition countries, especially in labor-intensive sectors. Small farmers are commonly the most susceptible agricultural producers. When institutional structures are incomplete or captured, small farmers in EECA are at risk of being excluded from value chains[6, 7, 8].

This study aims to show how small and medium-sized food processors establish efficient procurement systems in underdeveloped institutional environments of EECA by connecting with local farmers and incorporating them into national and international value chains. The article explores characteristics that influence connections between small farmers and processors to develop policies and support alternatives that can enhance group-based relationships between producers and processors.

We will analyze several sources to study smallholder-buyer business models that aim to integrate small farmers into marketplaces. The Food and Agriculture Organization (FAO) of the United Nations has combined the conclusions of five nation reports based on case studies from Azerbaijan, Kyrgyzstan, Serbia, Turkey, and Ukraine regarding the integration of small farmers into value chains led by processors[1].

The chapter is structured as follows. The next section presents a literature review on the incorporation of small-scale farmers into value chains. Foreign investments are crucial in this scenario. Hence, they merit a distinct subsection. A separate part focuses on cooperatives, which are viewed as a conventional tool for incorporating farmers into value chains. The next part outlines the favorable experiences of the case studies from five EECA countries. The chapter concludes with recommendations on how to promote the integration of small farmers into value chains through processor-driven initiatives.

2. INTEGRATION OF SMALL FARMERS INTO FOOD VALUE CHAINS IN TRANSITION COUNTRIES

2.1. Means of Integration

The transition process has caused a decline of agricultural output and decapitalization of the agricultural production system in Central and Eastern Europe [10]. As a result, the food and agricultural commodity value chains have undergone a tremendous change in the last decades [11]. The breakdown of the interactions among food chain participants—from farmers to retailers—resulted in supply disruptions of supply and deterioration of the quality of food products. Concurrently, changing consumer demands and the entry of foreign retailers and processors have called for groundbreaking reforms of food chains.

Transformation of the retail sector from state-run shops and open bazaars to “more modern”, large format retailers was stipulated by significant inflows of foreign investments. Their rapid development notwithstanding, the new retail formats have been expanding differently in different transition countries. This gradual expansion is often referred to as the “retail waves” concept [12]. In this context, the “first wave” countries’ supermarket sector grew from about 5% in the mid-1990s to 50% by the mid-2000s. These shares of modern retail formats are observable in Hungary, Poland and the Czech Republic. “Second wave” countries such as Bulgaria and Croatia are characterized by a 20–30% share of supermarkets and similar retail formats. “Third wave” countries, in turn, are those whose supermarket sector did not exceed 5% of total retail turnover in the mid-2000s, for example, Russia and Ukraine. Regardless of the market shares of modern retailers, existing evidence suggests that foreign companies are among the leaders in the retail sector of all Central and East-European countries. For example, the German Metro Group is the second largest retailer in Russia and Ukraine.

Foreign retailers and investors 'export' their business models. In the process of internationalization they are taking their own business models into the new markets [13, 14, 15]. Thus, modern management concepts and requirements toward business partners are exported. In the retail sector, this results in the following changes. The centralized, large and modern distribution centers and external specialized logistic firms substitute for traditional, local, store-by-store procurement. Furthermore, modern retailers set their own private standards of food quality and safety that are often much higher than those of national governments [12]. Some informed commentators suggest that these private standardization initiatives will further proliferate as the share of modern retailers is growing [16].

Profound structural change is expected in the agri-food sector due to foreign investment. Within new procurement systems, suppliers have to guarantee both timeliness and quality of delivered products. This leads to improvement of operations among domestic suppliers as they face fierce competition on the part of importers [17]. However, in the long run, foreign retailers aim to source up to 90% of their supplies from local producers [3, 4].

The observed process of synchronization of successive stages in the marketing channel from producers to consumers generally refers to vertical coordination. It does not include transactions on spot markets, where the commodity exchange is based on a price agreement only. It includes both vertical integration and productive partnerships (contracting) [18]. Assumingly, the higher the priority to secure quality and/or quantity of raw materials, the stronger the shift from spot market transactions toward advanced vertical coordination mechanisms.

Vertical coordination, introduced by modern procurement systems, favors (in theory) large scale farming. There are two reasons for this perspective to prevail. First, the complexity of the system is reduced and, thus, transaction costs are lower if there are a small number of large suppliers. Second, larger farms are less costly than smaller farms with regard to assistance provision.

However, small farmers are essential for ensuring the required supply quantities in most EECA countries. Therefore, retailers and processors are (still) compelled to include small farmers in supply chains and provide assistance to them based on the establishment of productive partnerships. Particularly in labor intensive sectors, small-scale farming has important cost advantages.

As generally regarded, productive partnerships among firms are supposed to be based on common goals, shared knowledge and pooled resources that altogether lead to better supply chain performance. The design of such relationships may vary from loose agreements to long-term and trust-based contracts [19].

The key for productive partnerships is contracting. In the context of agri-food supply chains, the main motivations of farmers to engage into contracting are the following: (1) income stability (to reduce risk compared to other ways of selling on traditional marketing channels); (2) improved efficiency (management decisions are transferred to the farmers); (3) market security (entering the contract provides a certain security in that the product will be sold if it meets with the requirements); (4) access to capital (contractor often provides inputs for farmers, which reduces the usage of credits) [20]. Food processors enter into contracts because they obtain control over input supply. Further, processors use contracts in order to achieve uniformity and predictability to suit consumers while they also benefit from lower costs in processing, packing and grading [21, 22, 23].

In general, one can distinguish between two types of contracts: marketing and production contracts. Marketing contracts address the issue of supply disruptions by private contractual initiatives [24, 25] whereas production contracts address quality concerns [26]. Additionally, these contracts are different with regard to the degree of control allocated and risk transferred across stages. In producer contracts, the contractor engages in the producer's decisions to a much higher extent than in the marketing contracts. This engagement can even expand toward ownership of critical production inputs whereas the role of marketing contracts is essentially in providing a market for the producer's goods [27].

Empirical evidence indicates that both types of contracts persist in EECA countries and their use is contingent upon the degree of market development. The less a market and its institutional environment are developed, the less likely it is that a complex system of vertical coordination will emerge and, this, marketing contracts will dominate. A more developed market, characterized by greater demand for higher quality products, entails a higher degree of vertical coordination with the wide use of production contracts. However, the use of marketing contracts is more reasonable if the higher quality products become standardized. The application of production contracts will then mainly pertain to consumer segments characterized by differentiated demands.

In practical terms, contracts that are used in EECA include farm management assistance, extension services, quality controls, farm input assistance programs, trade credit and even bank loan guarantees [1]. Thus, evidence suggests that the key actors (retailers and processors) find themselves constrained not by their own capital capacity but by that of other participants along the chains on which they depend for critical inputs. For the most part, this takes place because traditional lending institutions such as commercial banks do not give credit to enhance the interfirm product flow. In fact, this has been found to cause frequent contract breaches in EECA as farmers were not able to access basic production factors to fulfill a contract [25].

In addition, contract enforcement may be an important problem since public enforcement institutions are weak. Informal enforcement mechanisms such as social pressure, unacceptance of distrust, etc.) are also missing due to poor social capital. Therefore, farm assistance programs must be accompanied by appropriate governance mechanisms.

Thus, the search for quality is a key engine of vertical coordination, but what happens when the desired quality level is reached? Quality is becoming less of a driver while the need to enhance efficiency arises as the main motivation for vertical coordination. For example, in supply chains that bear high costs, retailers and processors work closely with their suppliers to reduce costs. Quality remains a key driver only when a higher than average quality is explicitly demanded by the customers or when it can be used for differentiation from competitors.

2.2. The Role of Foreign Direct Investments

Foreign direct investment (FDI) became an increasingly important element in global economic development and integration during the 1990s [28]. Ahrend names two general factors that make companies open subsidiaries abroad [29]. On the one hand, companies strive to sell more goods and services that they produce in their home countries. On the other hand, they want to launch production in a foreign country that would further enable sales to local and export markets. In essence, agri-food companies internationalize because of the same general reasons [30, 31].

Literature on the influence of FDI on transition economies mentions several positive effects of FDI. A number of authors agree that FDI facilitate economic growth and reduce poverty [32, 33, 34]. Several studies offer empirical evidence of the importance of FDI flows for economic growth in developing countries [35, 36]. Other advantages of FDI include technology transfer and technical innovation as well as enterprise restructuring [32, 37].

In the EECA countries, FDI induced the following major shifts in procurement systems: (1) procurement systems became largely centralized and based on large and modern distribution centers instead of local store-by-store procurement; (2) procurement became regionalized across borders; (3) traditional brokers were replaced by specialized wholesalers; (4) the logistics market became dominated by global firms; (5) retailers established preferred supplier systems; and (6) private quality standards were introduced [12]. Altogether these changes marked a growing role of vertical coordination in agri-food chains of EECA countries [2]. Studies from Bulgaria, Moldova and Slovakia show that, due to stricter and higher quality norms, more vertical coordination is taking place [38, 26]. In the Russian food sector, the foreign-owned food processors have managed to become the major competitors of the domestic ones, in particular in the dairy sector.

2.3. Cooperatives as a Means to Integrate Smallholders

The quest for quality, that requires tight coordination of interdependent activities in value chains, calls for particular attention toward the role of cooperatives [39, 40]. Bijman and Muradian mention that international donors and NGOs have (re) discovered the importance of cooperatives for rural development in general and for strengthening smallholders' access to markets in particular [39].

Because small-size producers are the backbone of agriculture in Central and East-European countries, their resources as well as output toned to be pooled to achieve the demanded quantity of supplies. Horizontal cooperation among smallholders, thus, gains in importance [41]. However, cooperatives face hard times in transition countries. In the Soviet era, farmers have been 'forced' to join collective farms. Thus, today, collective action still has a bad reputation as it is associated with loss of private ownership and freedom [42]. Furthermore, during Soviet times, collective farms and processing enterprises have proven to be very inefficient and subject to soft budget constraints. An additional problem with cooperatives is that there is often a lack of trust and social capital among farmers and villagers so that collective action is hindered already at the initial stage. Thus, Gardner and Lerman conclude that the evidence for cooperatives in agricultural production is still unfavorable [43].

However, for marketing and supply cooperatives, they observe a more promising situation. One reason for this is that new forms of cooperatives have been recently established [44]. A good example is a Hungarian Morakert cooperative where product quality and professional marketing are the first priority. Today Morakert's sales to retailers account for about 90% of its domestic turnover. Morakert's success is based on four key factors. First, filter rules are applied to membership. Second, quality and quantity of products are strictly coordinated. Third, trust is an inevitable aspect of communication between members and management. Fourth, private contract enforcement is established [45]. Morakert's procurement system is centralized, maintained in one place and supported by a common IT system [46]. An own brand serves as another marketing and coordination mechanism at Morakert. This example demonstrates how some problems of post-socialist economies can be overcome.

3. PECULIARITIES OF SMALLHOLDER INTEGRATION IN VALUE CHAINS

The above presented literature review shows that particularly small farms and households face serious production constraints caused by factor market imperfections in EECA countries. They do not have access to finance, they experience difficulties to buy (high) standard quality inputs, and lack technical and managerial capacity. Thus, more and more contract schemes (marketing and/or production contracts) and outgrower schemes have been established often accompanied with the provision of quality inputs, new technologies, credit and extension services to the farmers [10]. The current section provides further evidence of privately driven vertical coordination derived from the FAO reports on case studies from five EECA countries—Azerbaijan, Kyrgyzstan, Serbia, Ukraine and Turkey.

3.1. Marketing Channels of Small Farmers and Households

These general findings have been observed in the reports of all five countries. Disruptions of the agri-food value chains resulted in a dualistic structure of agricultural production. One the one hand, there are large corporate farm businesses such as agroholdings or state enterprises while, on the other hand, there are smallholders that account for considerable share of agricultural production in a number of sub-sectors. For example, in Ukraine, fully vertically integrated enterprises that are often referred to as agroholdings have rapidly developed whereas the labor intensive sub-sectors such as horticulture are marked by domination of small farms and rural households in production [47].

According to all country reports, the role of agricultural sector in overall gross domestic product (GDP) is diminishing. This fact notwithstanding, agriculture is still among major employers and, thus, is closely intertwined with rural development. In particular, all reports emphasize the importance of small-scale farms and households not only for own consumption but also for market supplies and employment of family members. At the same time, income disparity between urban and rural areas becomes larger in favor of the former. Particularly for young people, urban areas are more promising with regard to their future careers. Consequently, outmigration from rural areas is a common phenomenon in all countries.

This outmigration affects negatively the market size for direct sales as the main marketing channel of smallholders. Moreover, as exemplified by the Serbian case, modern retailers take over the shares of smallholders by offering more fresh products, in particular fruit and vegetables. Retailers are also able to offer these products for a reasonable price and standardized quality, thus engaging into stiff competition against directly marketing farmers and households. As a result, the attractiveness of direct marketing is shrinking.

Opposite to the negative effect of migration on the direct marketing of farm products, indirect marketing channels rather profit from this development. Moving to urban areas people have to buy processed food and have to buy in retail outlets. Therefore, the importance of indirect marketing channels is already high and will grow in the future despite growing competition from large retailers.

3.2. Obstacles to Integration into Value Chains

Smallholders in EECA face two main obstacles to integration into value chains. On the one hand, they do not have sufficient volumes of production. On the other, they often lack the required quality. In order to combat the first obstacle, the solution would be to pool quantities. A classical way would be the establishment of cooperatives. However, the reports have been very clear mentioning that cooperatives are not successfully operating. The reasons include negative reputation due to the historical background of cooperatives, operational inefficiencies and top-down implementation of the cooperative ideas, as well as taxation and administrative disadvantages. Another way would be the horizontal informal collaboration among smallholders. However, all studies report of a low level of trust and social capital among rural population. For example, the Ukrainian case study on Navigator-Agro demonstrated that there was a complete absence of trust from small and medium farmers in the idea of collaboration. Moreover, the level of trust between each other was very limited among rural inhabitants. A third way, as exemplified by the Turkish case, would be to sell the products through middlemen but low prices make sales at the spot market more profitable.

Regarding the second obstacle—the quality issue—literature provides reach evidence that quality can be improved based on vertical coordination. However, the literature most often refers to the examples of foreign investors who have leapfrogged quality by collaborating with large corporate farms. Modern and particularly foreign-owned retail chains prefer buying the needed larger volumes from a limited number of suppliers, thus favoring mainly corporate farm businesses. The reports demonstrate that larger processors also favor larger suppliers.

The concentration ratio and market power of modern retail chains and processors are increasing in EECA. Overall, this development favors large suppliers, especially branded manufacturers that invest in the achievement of appropriate quality levels for their branded products. The reports also reveal that imports are playing an important role in satisfying the demand for 'higher' quality products. However, interestingly, imports are also a major driving force of the competition in the low price segment. For example, Chinese tomato paste imports are replacing local Kyrgyz products. The Ukrainian report shows that domestic processors have lost 10% of domestic market owing to stiff competition with foreign companies. Another threat for Ukrainian producers resides in poor quality of local products.

Another issue in is high competition from imports due to membership of all scrutinized countries to the World Trade Organization (WTO). For example, Turkish pasta producers often substitute cheaper imported wheat for locally produced one. To this end, it is often easier to source from one large supplier from abroad than from a multiplicity of small local producers.

In Ukraine, concentration is taking place not only on the retail and processor levels but also on the farm level. Furthermore, due to large volumes of imported raw and processed products (e.g. dairy products, fruits and pork meat), there is a high competition. The willingness to invest into the development of cooperation with small producers is rather low for private companies. Smallholders have a limited access to state support, investments and credit resources. The result is that they experience a lack of qualitative seeds, fertilizer and mechanical appliances. Additionally, they have a lack of reliable information about markets.

Noteworthy, agricultural policies also contribute to dualistic production structures as well as to the development of large scale farming businesses. Driven by considerations of food security and food safety, most of the programs promote large scale farming and processing. For example, food security is among major priorities of the government in Azerbaijan where the related reforms have been conducive to the development of vertically integrated holdings. The Serbian and Turkish cases are marked by limited and lagged access of small farmers and processors to the information on relevant policy programs. In addition, the application for state support programs is often too complicated, thus constraining small producers or even precluding them from necessary support due to limited human and time resources. In all five countries, positive exceptions are the policies regarding storage facilities. Since small farmers and households face difficulties to store their harvest—particularly in the case of perishable products—policies are setting incentives to enlarge storage capacities. For example, in Azerbaijan, 28 cold storage facilities (and 17 cereal storage facilities) have been constructed with the help of the government in different regions of the country. A similar example can be found in the wine sector of Azerbaijan; a GIZ (Deutsche Gesellschaft für InternationaleZusammenarbeit) study has shown that particularly the rural infrastructure has to be developed in order to enhance the competitiveness of the sector and the inclusion of small wine growers [48].

The report on Kyrgyzstan demonstrated explicitly that even if successful cooperation is established, the optimal mode of transactions is also contingent upon the institutional setting. About 75% of farmland in Kyrgyzstan is under operations of small producers who face both production and marketing problems. The main issue is the lack of a stable buyer of the products. International technical assistance projects were temporarily helpful in achieving tighter cooperation between farmers and processors but the latter have reported that dishonest behavior of small farmers made such cooperation unsustainable. Despite advanced payments, fertilizer and seeds received from processors, farmers did not fulfill their commitments after harvesting. Instead, they sold their products at the fresh markets where prices were higher. Therefore, almost all processors prefer to operate through intermediaries or procurers that collect produce from the small-scale farmers and sell in bulk to the enterprises. Intermediaries and farmers predominantly collaborate based on short-term oral arrangements without signing any contracts. Given such circumstances, it is also no wonder that the relations between processors and farmers have worsened over time and spot market transactions have become the best choice. To this effect, a shift of export policy from unprocessed to processed foods made spot market transactions more favorable, thus decreasing the extent of vertical coordination in the sector. However, in the long run, most processors would like to work with cooperatives and larger suppliers rather than dealing with small-scale farmers.

In addition to institutional settings and opportunistic behavior, other obstacles have to be overcome to establish tight vertical linkages in supply chains. Low social capital and lack of trust are particularly constraining. As the case study on the Ukrainian Shyroke marketing group exemplified, processors had to provide some kind of collateral to suppliers prior to the start of cooperation in order to convince the smallholders of the reliability of intentions. A well-known game-theoretic procedure named hostage exchange is the case in point here [49].

Overall, it is evident that SME processors in all analyzed countries have to cope with very intense competition and a number of other problems, thus facing the need for clear corporate and marketing strategies. At the same time, as the report on Kyrgyzstan exemplifies, many SMEs still do not dispose of such strategies. A survey revealed that only a minor share of enterprises has strategic development plans. The managers of most companies neither participate in management trainings nor invest in capacity building of their staff. Even more threatening is the absence of customer orientation among SME processors in almost all analyzed countries.

3.3. Processors as Drivers of Integration

Small-scale agricultural producers in EECA nations are invited to collaborate on a smaller level compared to developed countries. If one takes into account the potential contribution of tiny and medium-sized processors, the situation can be altered. They are not appealing commercial partners for huge corporate farms, hence they are more willing to collaborate with small farms and households. The nation reports showcased how SME processors encourage collaboration among farmers in informal organizations. Collaborating with small and medium-sized processors is a promising strategy to incorporate smallholders into value chains.

Strong partnerships between processors and smallholders are mutually beneficial. Processors are assured of the quality and quantity of their raw input supply. Small-scale farmers and households value a dependable and secure outlet to sell their goods as their primary benefit. Smallholders prioritize secure market access over fair prices in their cooperation, according to all reports in this context. Vertical coordination is considered a key requirement for the successful establishment of sustainable value chains.

The case studies highlight successful instances of value chains that demonstrate client focus, possess a defined plan, and execute well. Successful SME processors do not attempt to compete with national or global cost leaders since they are unable to achieve the necessary economies of size and scope, making a cost leadership strategy unattainable. Successful examples involve utilizing a combination of niche and differentiation tactics to provide high-quality traditional items. Similar to the Serbian Zdravo Organic, distinctiveness is achieved through production-related features. Initial efforts are made to establish authentic brands, focusing on local, national, and international markets. Customer orientation extends beyond end consumers to encompass distributors, retailers, exporters, importers (e.g. Schwabe business), investors from other sectors (e.g. Zdravo Organic), overseas investors, and benefactors.

Due to rising competition in the retail industry, small and medium-sized enterprise processors are gradually being marginalized from this distribution channel. Sirela, a Serbian dairy company, responded to this growth by vertically integrating and establishing its own retail network. Currently, Sirela sells all its products through approximately 40 of its own stores, while other food manufacturers have copied similar approach and are establishing their own retail stores.

SME processors must adhere to market demands for consistent product volumes and excellent quality, regardless of their strategic orientations. Effective connections between the SME processors and their suppliers, such as small farms and homes, are crucial for success in this situation. Furthermore, processors should take the lead in initiating this collaboration. The case studies show that enduring and reliable vertical relationships exist, but smallholders are often seen as challenging partners who tend to act opportunistically by choosing to consume their products at home rather than selling them in the market.

Processors utilize outgrower programs specifically to enhance the quality of the products they get. They supply essential resources such as high-quality seeds, irrigation equipment, or specialized breed pigs, along with training possibilities like establishing a demonstration farm and extension services. Furthermore, processors may provide sophisticated financing options for farm supplies, as indicated in multiple papers. However, it is important to prevent misbehavior by farmers through contracts and fines, as shown by the successful example of the Agroplast cooperative in Kyrgyzstan that makes tomato paste.

3.4. Positive Experiences of Tight Vertical Linkages

Two common tactics are seen in creating effective vertical connections. The initial priority is to focus entirely on small and medium-sized family farms and dedicate resources to this form of collaboration. The second strategy is partial vertical integration, which includes acquiring or constructing farms while also collaborating with small farms simultaneously.

Exemplary instances of these two tactics can be observed in the dairy industry of Serbia. Sirela is one dairy company while Lazar is another. Retail chains are gaining more market strength in the Serbian dairy industry. Sirela and Lazar have chosen to establish their own retail networks in reaction to this development. However, the only commonality between the two organizations is their sourcing techniques, which are completely distinct. Lazar has effectively implemented vertical integration at a dairy farm with 500 milk cows. Sirela has collaborated with several small farms due to their negative experience with full vertical integration in primary production. Sirela provides a secure sales market, reliable payments, enhanced access to subsidies, extension services, complimentary use of milk coolers, and on-demand assistance to farmers.

Successful collaborations between SME processors and small family farmers can be seen in the fruits, vegetables, and grain sectors. Many SMEs focus on the home market, but some, like Zdravo Organic and Vitamin in Serbia, are heavily engaged in exporting. SME food processors gain from collaborating with small farmers by securing input markets. Processors supply farms with inputs like seeds, fertilizer, chemicals, irrigation equipment, and advance payments. Extension service is also available.

The Ukrainian case studies have demonstrated how processors and small farms can build mutually beneficial collaboration. The rural population in Ukraine has low levels of social capital and cooperativeness, which hinders joint efforts by processors or wholesalers. Nevertheless, this challenge can be surmounted by employing leadership techniques. Processors have started forming farmer organizations under the guidance of both formal and informal rural community leaders, which has increased the trust among these groups. Moreover, processors offer financial assistance to these groups at the start of their collaboration. This also enhances farmers' confidence in a processor. Navigator-Agrofirm invested in building a small logistic center in a community.

The Ukrainian case of the Kolos firm showed that collaboration between small-scale farmers and big agricultural conglomerate can be achieved. Kolos, a meat-processing company from Chernivtsi, collaborated with Kamyanets-Podilsky University to create a breeding and reproduction method for a Dutch pig breed, which was subsequently shared with small-scale farmers. The primary goal of this effort was to expand the supplier base and maximize the utilization of Kolos's industrial facilities. The agricultural aid initiative involved on-site demonstrations and training for suppliers at an experimental farm, along with the implementation of a quality monitoring system at the rural family level. The bundle of supplementary services provided to farmers consisted of piglets, feed, veterinary care, and financial assistance. As an early investment in trustful cooperation, the processor has enabled farmers to keep one or two pigs for their personal consumption. However, there were instances where farmers attempted to deceive by augmenting the weight of delivered pigs through undisclosed ways. These examples demonstrated the necessity of including penalties in contracts to deter opportunistic behavior and dissatisfaction from the honest supplier.

The Kyrgyz report showcases two instances of successful integration of smallholders pushed by processors. Agroplast cooperative supplies farmers with seeds and fertilizer to enhance the quality and quantity of their crops. Furthermore, farmers receive training once a year. At the start of each year, the cooperative summons all farmers to strategize output quantities and ascertain the agricultural input requirements of farmers (such as seeds, fertilizers, etc.). Agroplast also offers prepayment options. Both parties view their relationship as trustworthy, but they are aware that a single error, such as fraud or dishonesty, could quickly erode the existing trust.

In addition to making and keeping agreements, successful cooperation has also been driven by the establishment of informal supplier groupings. Agroplast has established multiple farmer groups, each overseen by a reputable leader. This individual serves as a liaison between Agroplast and the farmers. The group leaders oversee information exchange, coordinate logistics, and manage delivery schedules. Group leaders also serve as guarantors for group members seeking financial assistance from Agroplast.

The second case study was carried out at Galanfarm, a company that distributes valerian to a German pharmaceutical company. Galanfarm invited small farmers to participate in a Farmer Field School funded by a German donor and operated in collaboration with a local consulting agency to improve the quality of valerian production. In addition to receiving thorough training on all crucial production processes, farmers were given the chance to create informal groups and choose leaders within those organizations. Just like in the Agroplast situation, these group leaders act as intermediaries between Galanfarm and the farmers. They manage the delivery quantities and schedules, and ensure the quality of the raw materials provided. This story highlights the significant impact that international donors can have on the advancement of procurement systems and the inclusion of smallholders in them.

4. CONCISE SUMMARY

Our analysis of five nation reports shows that small-scale producers continue to play a crucial role in agricultural value chains in transitioning economies, especially in the production of perishable and labor-intensive goods. Marketing channels for items from small-scale producers are varied and encompass direct sales to rural communities, processing sector, wholesale, food retail, and exports. Simultaneously, personal consumption holds significant importance, particularly during periods of low market pricing. The future of small farmers relies on the advancement of the processing industry as a whole, especially small and medium-sized enterprise processors. Therefore, policy should strive to establish a conducive climate for SMEs.

The smallholder involvement in value chains should aim to develop customer focus. Country reports consistently indicate that competition is increasing as domestic markets experience an influx of imported goods. Concentration processes are occurring in the downstream levels of the value chain in the nation studies reviewed. Small manufacturers must adhere to the demands of both their business clients and final consumers in order to effectively address these two trends.

The five findings unequivocally show that a cost leadership strategy is not suitable for SMEs. Cost leadership necessitates producing big volumes to achieve efficiency at scale. Therefore, at the national level, major local processors are more capable of executing this plan. Cheaper imports increase competitiveness within the framework of the WTO. This also applies to exports. Therefore, tiny manufacturers must implement a differentiation strategy to achieve long-term success. Possible methods for distinguishing include improving quality, obtaining certification, creating a strong brand, producing traditional or local products, and more.

Within vertical coordination, the case studies have highlighted several key success criteria. Initially, providing input help from processors to farmers is essential for farming. Vitamin, a processing company, has implemented a contract farming program for cultivating pepper seeds. This program is expanded by distributing seeds to a bigger number of farmers who cultivate the peppers. Seed distribution is accompanied by the provision of additional inputs and assets like fertilizer and irrigation equipment. Processors supply animal breeds from Ukraine and milk cooling tanks from Serbia to their suppliers. Suppliers are provided with precise production instructions and the processors closely monitor the production process in addition to providing input. Noncompliant farmers face penalties.

Additionally, downstream partners offer cash support to farmers, in addition to input supply. This help encompasses several forms of money, including loans, forward payments, and promissory notes. Prior to the commencement of the relationship, funds were allocated towards a modest logistics facility. Subsidies are provided to smallholders in certain countries only when a processor requests them on behalf of the smallholders.

Thirdly, extension services are offered. Processors authorize their workers and engage external specialists, establish demonstration farms, and collaborate with local institutions, foreign donor organizations, and technical support organizations to provide training for the farmers.

Furthermore, the establishment of formal and informal groups of farmers fosters collective activity and trust. Through training in these groups, farmers gain confidence in value chain partners, exercise social control, meet production quotas, and diversify production risk. It is crucial for groups of farmers to choose group leaders who are respected individuals. Group leaders act as a cohesive force that unites all group members and contribute voluntary, grassroots elements to this form of collaboration that was originally supported by external processors. Therefore, all group members share the same goal and anticipated advantages. All reports indicate that processors are in favor of replacing intermediaries by collective action, such as establishing official farmer associations. Cooperatives, in their conventional form, are not prevalent in EECA countries because to their association with collectivization and socialist top-down methods, which have been shown to be ineffective and corrupt. One option is to create cooperative frameworks inspired by the concepts of new generation cooperatives. Avoid using the word "cooperative."

Overall, the key factor in incorporating smallholders into value chains is ensuring the creation of sustainable advantages for them. It is important to show the importance of reliable and secure marketing channels with equitable payment systems in this situation.

5. AT A GLANCE

The economic collapse during the initial stages of the transition significantly impacted food supply dynamics in Eastern European and Central Asian (EECA) nations. Small and medium-sized farm suppliers and processors had a shortage of essential production inputs, while processors and retailers encountered issues with inadequate quantity and quality of supply. Simultaneously, shifts in consumer preferences and the influx of international investors in the retail and processing industries required substantial and prolonged modifications and adaptations in the organization of food supply chains to address these issues. The current chapter illustrates how small and medium-sized food processors establish efficient procurement systems in challenging institutional environments of EECA through an in-depth literature review and analysis of five case studies commissioned by the Food and Agriculture Organization (FAO) of the United Nations. The chapter explores the elements that influence the connections between small farmers and processors, and their incorporation into global value chains to create support and aid strategies.

REFERENCES

1. Swinnen JFM. When the Market Comes to You—Or Not; the Dynamics of Vertical Coordination in Agri-Food Chains in Transition. Final Report on Dynamics of Vertical Coordination in ECA Agri-Food Chains: Implications for Policy and Bank Operations. Washington, DC: The World Bank; 2005
2. Swinnen JFM, Dries L, Noev N, Germenji E. Foreign Investment, Supermarkets, and the Restructuring of Supply Chains: Evidence from Eastern European Dairy Sectors. LICOS Discussion Papers: Leuven; 2006
3. Hanf JH, Gagalyuk T. Supply chain quality and its managerial challenges—Insights from Ukrainian agri-food business. Journal of East-European Management Studies. 2009;14(4):332-356
4. Gagalyuk T, Hanf JH, Hingley M. Firm and whole chain success: Network management in the Ukrainian food industry. Journal on Chain and Network Science. 2013;13(1):47-70
5. Hanf JH. Processor driven integration of small-scale farmers into value chains in Eastern Europe and Central Asia. In: Tanic S, editor. Enhancing Efficiency and Inclusiveness of Agri-Food Chains in Eastern Europe and Central Asia. Budapest: FAO Regional Office for Europe and Central Asia; 2015. pp. 13-36
6. Mamonova N. Resistance or adaptation? Ukrainian peasants' responses to large-scale land acquisitions. The Journal of Peasant Studies. 2015;42(3-4):607-634
7. Gagalyuk T, Schaft F. Corporate Social Reponsibility in Agribusiness. Kyiv: German-Ukrainian Agricultural Policy Dialogue; 2016
8. Dorobantu S, Kaul A, Zelner B. Nonmarket strategy research through the lens of new institutional economics: An integrative review and future directions. Strategic Management Journal. 2017;38:114-140

9. Hanf JH. Processor Driven Integration of Small Farmers into Value Chains in Eastern Europe and Central Asia (EECA) as an Alternative to Cooperatives—A Synthesis Paper. Budapest: FAO Regional Office for Europe and Central Asia; 2014
10. Dries L, Germenji E, Noev N, Swinnen JFM. Farmers, vertical coordination, and the restructuring of dairy supply chains in central and Eastern Europe. World Development. 2009;37(11):1742-1758
11. Swinnen JFM, Maertens M. Globalization, privatization, and vertical coordination in food value chains in developing and transition countries. In: Proceedings of the 26th Conference of International Association of Agricultural Economists; 12-19 August 2006; Gold Coast, Australia
12. Dries L, Reardon T, Swinnen JFM. The rapid rise of supermarkets in central and Eastern Europe: Implications for the agrifood sector and rural development. Development Policy Review. 2004;22(5):1-32
13. Hanf JH, Pieniadz A. Quality management in supply chain networks—The case of Poland. International Food and Agribusiness Management Review. 2007;10:102-128
14. Palmer M. Multinational retailer expansion: Learning from Tesco's experience. International Journal of Retail and Distribution Management. 2005;33(1):23-48
15. Roberts GH. Auchan's entry into Russia: Prospects and research implications. International Journal of Retail and Distribution Management. 2005;33(1):49-68
16. Fulponi L. Private voluntary standards in the food system: The perspective of major food retailers in OECD countries. Food Policy. 2006;31(1):1-13
17. van Berkum S. Restructuring and vertical coordination in the dairy sector in Romania. In: Swinnen JFM, editor. Case Studies: The Dynamics of Vertical Coordination in Agrifood Chains in Eastern Europe and Central Asia. Washington, DC: World Bank; 2005. pp. 167-189
18. Swinnen JFM, Maertens M. Globalization, privatization, and vertical coordination in food value chains in developing and transition countries. Agricultural Economics. 2007;37(1):89-102
19. World Bank. Rural Finance Innovations: Topics and Case Studies. Washington DC: The World Bank; 2005
20. USDA. Farmers' Use of Marketing and Production Contracts, Report No. 747. Washington, DC: Farm Business Economics Branch of the U.S. Department of Agriculture, Rural Economy Division, Economic Research Service; 1996
21. Boland M, Barton D, Domine M. Economic Issues with Vertical Coordination. Agricultural Marketing Recource Center: Ames, IA; 2002

22. Sykuta M, Parcell J. Contract structure and design in identity-preserved soybean production. Review of Agricultural Economics. 2003;25(2):332-350
23. Tsoulouhas T, Vukina T. Integrator contracts with many agents and bankruptcy. American Journal of Agricultural Economics. 1999;81:61-74
24. Dries L, Swinnen JFM. Globalisation, quality management and vertical coordination in food chains of transition countries. In: Proceedings of the 92nd EAAE seminar on Quality Management and Quality Assurance in Food Chains; 2-4 March; Göttingen
25. Gow HR, Swinnen JFM. Up- and downstream restructuring, foreign direct investment, and hold-up problems in agricultural transition. European Review of Agricultural Economics. 1998;25:331-350
26. Gorton M, White J, Chernyshova S, Skripnik A, Vinichenko T, Dumitrasco M, Soltan G. The reconfiguration of post-soviet food industries: Evidence from Ukraine and Moldova. Agribusiness: An International Journal. 2003;19:409-423
27. Martinez SW, Reed A. Vertical Coordination by Food Firms Rising Along with Contract Production, Report AIB-720. Washington, DC: Economic Research Service/USDA; 1996
28. UNCTAD. World Investment Report 1999-2003. Geneva: United Nations Conference on Trade and Development (UNCTAD); 2003
29. Ahrend R. Foreign Direct Investment into Russia—Pain without Gain? A Survey of Foreign Direct Investors. Russian-European Center for Economic Policy; 2000
30. Stange H. Die InternationalisierunglandwirtschaftlicherUnternehmen [PhD thesis]. Halle: IAMO; 2010
31. Pall Z, Hanf JH. A multi-perspective analysis of food retail internationalization—Insights from foreign retailers on the development of the Hungarian and eastern European markets. Management & Marketing Challenges for the Knowledge Society. 2014;8(4):593-606
32. Barrell R, Holland D. Foreign direct investment and Enterprise restructuring in Central Europe. Economics of Transition. 2000;82:477-504
33. Bevan AA, Estrin S. The determinants of foreign direct investment into European transition economies. Journal of Comparative Economics. 2004;32:775-787
34. Broadman HG, Recanatini F. Where Has All the Foreign Investment Gone in Russia? Washington DC: The World Bank; 2001
35. Blomström M, Sjöholm F. Technology transfer and spillovers: Does local participation with multinationals matter? European Economic Review. 1999;43:915-923

36. Borensztein E, De Gregorio J, Lee JW. How does foreign direct investment affect economic growth? Journal of International Economics. 1998;45:115-135

37. Hooley GJ, Cox AJ, Beracs J, Fonfara K, Snoj B. The role of foreign direct investment in the transition process in Central and Eastern Europe. In: Hooley G, Loveridge R, Wilson D, editors. Internationalization: Process, Context and Markets. Academy of International Business Series. New York: St. Martin's Press; London: Macmillan Press; 1998. pp. 176-200

38. Dries L, Swinnen JFM. Foreign direct investment, vertical integration, and local suppliers: Evidence from the polish dairy sector. World Development. 2004;32(9):1525-1544

39. Bijman J, Muradian R. Cechina agricultural cooperatives and value chain coordination: Towards an integrated theoretical framework. In: Helmsing B, Vellema S, editors. Value Chains, Inclusion and Endogenous Development: Contrasting Theories and Realities. New York: Routledge; 2011. pp. 82-101

40. Shepherd AW. Approaches to linking producers to markets. In: Agricultural Management, Marketing, Management and Finance Occasional Paper No 113. Rome: FAO; 2007

41. Hanf JH. Challenges of a vertical co-ordinated agri-food business for co-operatives. Journal of Co-operative Studies. 2009;42(2):5-13

42. Török T, Hanf JH, Gruzina Z. Co-operatives in the Latvian agri-food business—Agents of change? Journal of Co-operative Studies. 2010;43(2):16-23

43. Gardner B, Lerman Z. Agricultural cooperative enterprises in the transition from socialist collective farming. Journal of Rural Cooperation. 2006;34(1):1-18

44. Hanf JH, Török T. Co-ops as a way to integrate small farmers in supply chain networks? Journal of Rural Co-operation. 2009;37(1):20-31

45. Bakucs LZ, Fertő I, Szabó GG. Innovative Practice Hungary: Morakert Cooperative. Budapest: Institute of Economics. Hungarian Academy of Sciences; 2007

46. Fertő I, Szabó GG. Vertical Co-ordanitaion in Transition Agriculture: A Hungarian Cooperative Case Study, KTK/IE Discussion Paper. Budapest: Institute of Economics, Hungarian Academy of Sciences; 2002

47. Gagalyuk T. Strategic role of corporate transparency: The case of Ukrainian agroholdings. International Food and Agribusiness Management Review. 2017;20(2):257-277

48. Hanf JH. Assessment of Competitiveness and Potential for Export of Azerbaijani Wines to the EU Market. Private Sector Development Programme South Caucasus. Bonn/Eschborn: Deutsche Gesellschaft für InternationaleZusammenarbeit (GIZ) GmbH; 2014

49. Axelrod R, Hamilton WD. The evolution of cooperation. Science. 1981;211(4489):1390-1396

11 ENHANCED ECONOMIC BENEFITS RESULTING FROM CLOSER INTEGRATION OF AGRI-BUSINESS AND COAL SEAM GAS INDUSTRIES

1. INTRODUCTION: AT A GLANCE

The rising worry about climate change has heightened environmental consciousness and spurred an international effort for countries to decrease their carbon footprint by enacting measures to diminish greenhouse gas emissions [1, 2, 3], as exemplified by the Paris Climate Change Conference and the subsequent COP21 agreement. Australia is focusing on transitioning to a lower carbon economy by developing Coal Seam Gas (CSG) and other unconventional natural gas supplies as part of global energy policy measures. [4]. Australia has an abundant supply of CSG resources estimated to be around 168,600 Petajoules (PJ) in 2012 [5], with potentially rich CSG areas, yet to be explored [6]. With growing demand for low-cost gas production, Queensland is playing a key role in Australia's CSG exports with the construction of three CSG to liquefied natural gas (LNG) facilities (each with two LNG trains), worth approximately ~ $USD60 billion [7, 8, 9] of infrastructure investment.Although there are economic benefits in supplying CSG-sourced energy to both local and international markets and gas-fired electricity has lower emissions compared to coal, there is public concern regarding the growth and progress of Australia's CSG business. The concerns usually involve environmental difficulties such as managing the massive amounts of saline-rich CSG associated water produced during CSG extraction and conflicts over land use with existing agricultural enterprisesAlthough there are economic benefits in supplying CSG-sourced energy to both local and international markets and gas-fired electricity has lower emissions compared to coal, there is public concern regarding the growth and progress of Australia's CSG business. The concerns usually involve environmental difficulties such as managing the massive amounts of saline-rich CSG associated water produced during CSG extraction and conflicts over land use with existing agricultural enterprises. [4]. Due to the total dissolved solids (TDS) content of the CSGAW, it requires desalination and amendment to some degree before use in most surface applications. Recognising various industries or entities that can beneficially use CSG by-products will help alleviate the concern associated with the large brackish-water flows [4]. It is crucial to have carefully planned water management policies to strategically manage the volumes of water generated from the CSG industry and thereby maintain a sustainable balance [10, 11]. Careful evaluation of the beneficial use of the CSG-associated by-products can promote co-existence of other complementary industries that could provide sources of additional revenue for the CSG industry [5, 6] and expanded or new agribusiness opportunities. Potential end uses of the CSGAW that have been studied in this article include, irrigation, watering of livestock, abattoir/meat processing industry, leather industry/tanneries, discharge to surface waters, aquifer recharge, artificial lake or constructed wetlands for recreation and ecosystem diversity, coal mine water, cooling tower water, saline inland aquaculture, water storage to combat rural fires, and growth medium for cultivating microalgae for the biofuel industry [12, 13].

Australia's CSG business is often located in areas with strong CSG potential and limited urban growth. These regions are typically characterized by extensive farmland and fertile agricultural grounds used for livestock and irrigation. Hence, it would be advantageous to optimize prospects for coexistence between the CSG industry and agribusiness by fostering complementing industries that are already prevalent in rural economies based on agriculture. In this article, coexistence refers to the simultaneous operation of the CSG industry alongside the local API, usually located near CSG operations. APIs are businesses that support and contribute to the sustainability and growth of local agriculture-based supply chains. Because many CSG developments are located in rural areas near agricultural lands, it is advantageous to establish regional synergies to help boost economic development and growth by utilizing the skills and knowledge of the local workforce. It is crucial to collaborate and connect with local landowners and the agricultural community to ensure the successful creation and seamless integration of the CSG industry with current agribusinesses. This chapter examines the possibilities for coexistence with the existing agriculture industry, CSG industry, and API's by exploring the beneficial utilization of by-product CSGAW and services by other businesses. A coexistence model is proposed for integrating agribusiness with coal seam gas (CSG) activities in locations where CSG is abundant. This model can be implemented in existing CSG-rich areas or in new sites planned for CSG extraction and processing in agriculture-based regions, ensuring the continued relationship between food production and energy generation.

2. CSG BACKGROUND

2.1. Natural Gas from Unconventional Reservoirs

The composition of the extracted gas from both conventional and unconventional reservoirs is mostly methane [4, 14]; however, the source rock strata and the extraction techniques dictate the classification of the natural gas [8, 15]. Table 1 provides a summary of the main differences between natural gas sources. In the case of conventional gas, the natural gas migrates through buoyancy and natural pressure gradients within permeable strata (porous sandstone, siltstone or carbonate geological formations) to a point where it becomes trapped and therefore may not even require pumping to collect at the surface [15]. However, for unconventional natural gas (such as CSG), low permeability strata hold gas in place via capillarity or adsorption rather than buoyancy effects [14]. In Australia, the CSG industry is the most developed out of the remaining gas types sourced from unconventional reservoirs [5].

2.2. CSG Extraction

Coal seam gas (CSG) is an unconventionally sourced natural gas which usually contains approximately 95 + % methane, and is found adsorbed within the underground coal seams [4, 5, 8]. Coal is a carbonaceous or carbon-based sedimentary rock that formed from terrestrial organic matter such as trees, which decayed and compressed over many millions of years [16, 17]. Due to ongoing high pressure and temperature-associated compaction processes from the deposition of overlaying strata, the coal was naturally buried to varying depths depending on the extent of forces experienced by the associated geology [8, 16, 18]. This coal formation process is known as coalification [4, 19]. Depending on the geological history, the coal is classified into different ranks which are defined as the extent or level of coal maturation [8, 20, 21]. Low coal ranks are typically located close to the surface and are relatively 'younger' compared to higher coal ranks which have been buried deeper over longer time periods [8, 22]. With coalification associated processes, there could be thermogenic methane produced as a result of chemical reactions within the decaying organic matter and once generated, it becomes adsorbed into the matrix of the coal [23, 24]. Additionally, biogenic methane can also be produced from microbial activity, typically at temperatures less than 70° Celsius and at shallow depths. Biogenic methane can also be adsorbed into the coal matrix [25, 26]. Apart from methane, additional gases such as nitrogen and carbon dioxide also have the potential to migrate through the coal strata and consequently get adsorbed into the coal matrix in varying amounts [8, 27]. Geological investigation techniques and organic geochemistry analysis can reveal the most likely source and process from which the gas originated [20, 23].

Table 1. Some key differences between natural gas sources (adapted from [8]).

Attributes	Natural gas from conventional reservoirs	Natural gas from unconventional reservoirs		
		CSG	Shale gas	Tight gas
Hydrocarbon composition	Mainly methane (impurities ethane, propane, butane, condensate) May also be co-located with oil	Mainly methane (usually 95 + % purity)	Mainly methane but can have condensate	Mainly methane
Typical host rock permeability (TP)	Underground reservoir in sandstone, siltstone, or carbonate rock TP: ≥ 1 mD	Coal seams the coal matrix TP: 1–10 mD	Shale rock (more impermeable than coal) TP:10–3–10–9 mD	Varied rock locations (gas migrates into low permeability limestone & sandstone or siltstone reservoirs TP: 10–3–1 mD
Typical depth	1000–6000 m	200–1000 m	1000–2000 m	>1000 m
Extraction	Vertical/directionally drilled wells Gas transport due to natural pressure and buoyancy	Desorbed by depressurization of coal seam by water removal Vertical, horizontal or directionally drilled wells In Australia around 30–50% of wells will require stimulation in the form of hydraulic fracturing	Shale is highly impermeable and requires hydraulic fracturing Horizontally drilled wells	Vertical, horizontal or directionally drilled wells Large scale hydraulic fracturing required May need well acidizing to stimulate gas production from low permeability wells
Significant resource location	WA, SA, QLD & offshore (federal)	QLD, NSW	SA, NT, QLD & WA	WA, SA, VIC
Significant production location	WA, SA & offshore (federal)	QLD, NSW	SA	—

The geological structure of coal is characterised as a coal matrix (containing micropores), which is surrounded by a network of water-filled cleats or fractures [20]. Over time as the CSG resource is formed, it is adsorbed within the coal matrix and is typically found adjacent to the water-filled cleat structure [18, 20]. If the pressure in this coal-cleat system is decreased by drilling wells and producing water from the cleats, then the methane gas loses its adsorptive affinity with the coal structure and consequently de-sorbs and migrates with the water through the coal structure and is collected at the surface through production wells, as a gas-water mixture [8, 28, 29]. The CSG resource is then piped to a processing facility where it undergoes dehydration and compression. The gas is then transported via pipeline network to power stations for electricity generation [30]. Since the completion of the offshore LNG facilities, Queensland has been exporting liquefied CSG to international gas markets [31].

2.3. CSG Water Production

Generated in large volumes during the CSG extraction process, the CSGAW is regarded as one of the major by-products of the CSG production process, the other being salt (which is dissolved in the associated water) [32]. Figure 1 is a schematic representation of a generic CSG production curve for gas and CSGAW. Actual production curves are highly variable across a particular asset or between sedimentary basins. For example, the average CSG well in the Surat Basin in Queensland produces between 1 and 2 million standard cubic feet per day (MMscf/d) of gas but the best wells exceed 20 MMscf/d [9]. The historical ratio of water production to gas production across all of Queensland's CSG wells over time is plotted in Figure 2 and ranges between about 60 and 120 ML/PJ. Initially, when the CSG production wells are depressurized, large volumes of CSGAW are produced [4, 5]. As time progress, these significant water volumes decline, with increasing CSG flows [8, 32, 33]. Typically, the flow of CSG then gradually falls towards the end of the life of the CSG production well, when it can be decommissioned [34].

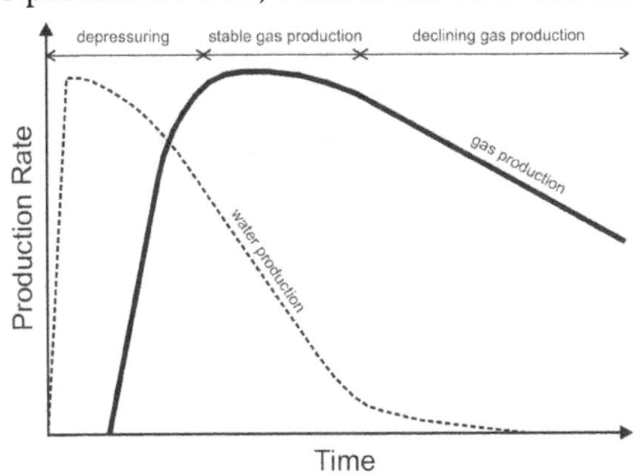

Figure 1. Typical CSG well production stages (modified from [14]).

2.3.1. Water Quality

CSG water chemistry is influenced by the geochemistry of the originating coal seams from which the water was removed from, as well as extent of interactions with other subsurface groundwater flows [8, 34]. CSGAW has been typically characterised with high levels

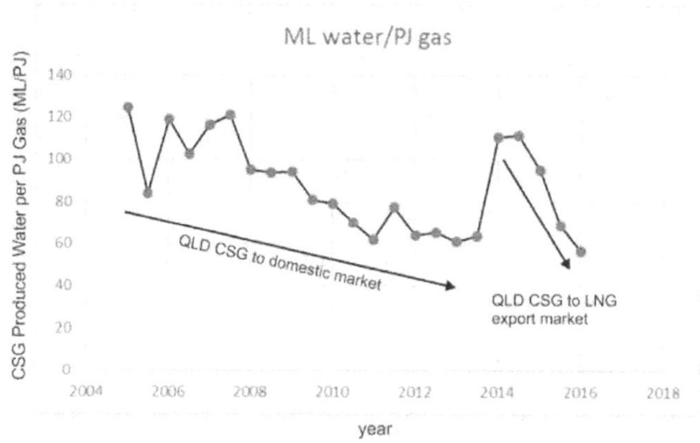

Figure 2. Queensland CSG production presented as Megalitre (ML) of produced water per Petajoule (PJ) of gas.

of dissolved solids & salts, oil based compounds (if thermogenic) and metals [3, 32, 35]. Chemicals used on the CSG operator's sites during well construction, drilling, stimulation and maintenance activities [8] may also be present in the chemical profile of the CSGAW. The characteristic quality of CSGAW is outlined in Table 2. CSGAW extracted from the Surat Basin has been typically characterised as being alkaline in nature, with high levels of sodium, bicarbonate and chloride content [36]. Water is classified as 'brackish water' when the TDS levels fall characteristically between 3000 and 15,000 milligrams per litre (mg/L); CSGAW is typically classified as 'brackish water' as it has total dissolved solids (TDS) ranging between 1200 and 7000 mg/L [6, 37]. Adequate water treatment and careful management practices are critical to prevent harmful effects on the environment and end user [38].

Table 2.Surat Basin CSGAW quality & acceptable livestock watering limits ([13, 36, 39]).

Water quality parameter	Unit	Range	Acceptable livestock watering limits
pH	—	8–9	None prescribed
Total dissolved solids (TDS)	mg/L	1200–7000	Table 6; Table 7
Sodium adsorption ratio (SAR)	—	107–116	None prescribed
Fluoride	mg/L	0.77–4.5	2–4
Sodium	mg/L	300–3461	None prescribed
Magnesium	mg/L	4–13	None prescribed

Water quality parameter	Unit	Range	Acceptable livestock watering limits
Silica	mg/L	19–51	None prescribed
Sulphate	mg/L	5–10	None prescribed
Chloride	mg/L	550–2092	None prescribed
Potassium	mg/L	20–78	None prescribed
Calcium	mg/L	2.3–24	None prescribed
Manganese	mg/L	0.07–0.10	None prescribed
Iron	mg/L	0.07–4.50	None prescribed
Bicarbonate (as CaCO3)	mg/L	580–2060	None prescribed

2.3.2. CSG Water Treatment Methods and Brine Management Options

The direct application of untreated CSGAWA is limited as its quality is often less than the required water quality of many end users [4, 40]. As previously mentioned, CSGAW contains levels of salt and other trace elements that may need to be removed before it is suitable for use. Therefore, most water treatment technologies rely on desalination methods such as reverse osmosis (RO) that then generate a highly concentrated saline effluent waste stream (brine) and a treated CSG water (permeate) stream [5, 6, 41, 42]. Many of the CSGAW treatment technologies are based on the idea of increasing the water recovery rate and consequently minimising the volume of brine [6, 38]. Furthermore, the viability of treatment processes is also largely determined by the cost factor associated with capital and operating expenditure [5].

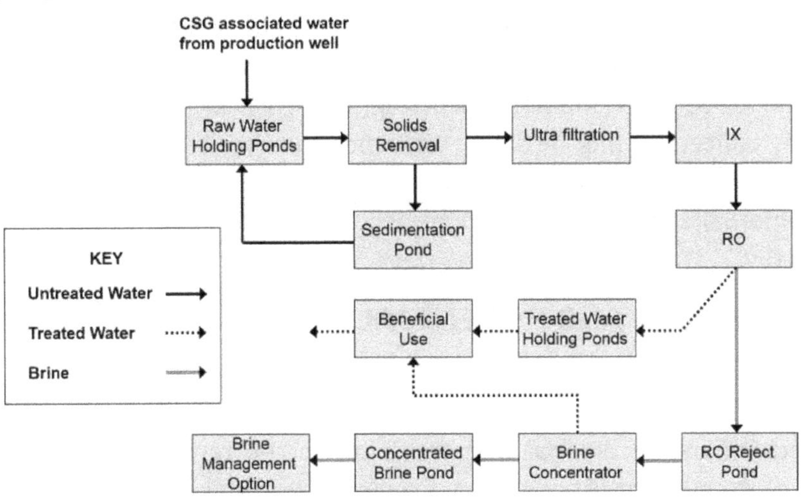

Figure 3. CSGAW processing scheme at Kenya Water Treatment Plant (modified from [39]).

For the RO plant to run efficiently there may be pre- and post-treatment required of the CSGAW. The major stages of CSGAW treatment include feed collection ponds (water collected to homogenisefeedstock), ultra-filtration (removal of particulate matter), ion exchange (IX) (reduction of water hardness ions, Ca2+ and Mg2+) and RO (desalination) units [6]. Chemical amendments and conditioning with dosing additives is further applied to ensure the treated CSGAW is suitable for the end user [5]. As an example, Figure 3 represents the overall CSGAW treatment process that is employed at the Kenya Water Treatment Plant operated by QGC Pty. Ltd. and managed by SunWater [43].

The saline waste effluent stream produced by the RO processing unit is typically further concentrated through the mechanical and thermal brine concentration units [6]. The brine concentration system is an integration of dehydration technology which includes, heat exchangers, falling film evaporation vessels, gas powered compressors, gas fired auxiliary heat chambers and de-aerators [44].

2.3.2.1. Brine Management Options

Brine is regarded as the concentrated saline effluent that is generated as the waste output stream from RO water treatment or brine concentrators [5]. Managing brine in an efficient and environmentally acceptable manner is of utmost importance to the CSG industry. One possible brine management option is to inject the brine generated from CSGAW treatment into a 'geologically isolated' containment that is at an adequate distance from any groundwater source. An alternative option is to evaporate the saline effluent (brine) to a more concentrated smaller volume or to further evaporate the brine to generate a dry solidified salt, which can then be transported to a waste disposal facility (operated by CSG company or off-site).

Therefore, an underlying aspect of brine management is brine volume reduction, to ease downstream processing of the large volumes of brine that will be generated over the life of CSG development. Growing interest is arising in minimising brine volumes by concentrating the saline effluent generated from RO water treatment, to ultimately produce commodity crystallised salts of potential commercial value. 'Recoverable Salts' include sodium bicarbonate, sodium carbonate and sodium chloride [4].

3. CSG SERVICES AND POTENTIAL AGRIBUSINESS PROMOTING INDUSTRIES

CSG developments primarily in Queensland's Surat and Bowen basins have introduced enhanced regional infrastructure to the remote landscape [45]. The presence of the CSG industry within a regional setting has introduced many new businesses that were not previously existent in the area. This has allowed for increased business activity and economic growth of the regional centres near the CSG industry [46]. Some of these enhanced community services facilitated by CSG developments are summarised in Figure 4.

Aside from community services, the establishment of the CSG industry in Australia has introduced an array of CSG field supporting infrastructure including underground gas and water gathering networks, gas processing facilities, water treatment plants, transportation networks & telecommunication systems to the CSG producing regional centres of Australia [4], many of which are on agricultural-rich lands [33]. The agriculture industry is by far the most established industry across a large part of Australia's regional area where many of the CSG developments are also located. Such an area is the agriculture-rich heartland of the Surat Basin in southern Queensland which is dominated by irrigation and cattle grazing lands [33]. An example of a regional setting that has experienced resource expansion such as this, is the Western Downs region in Queensland, due to its location within the CSG producing Surat Basin.

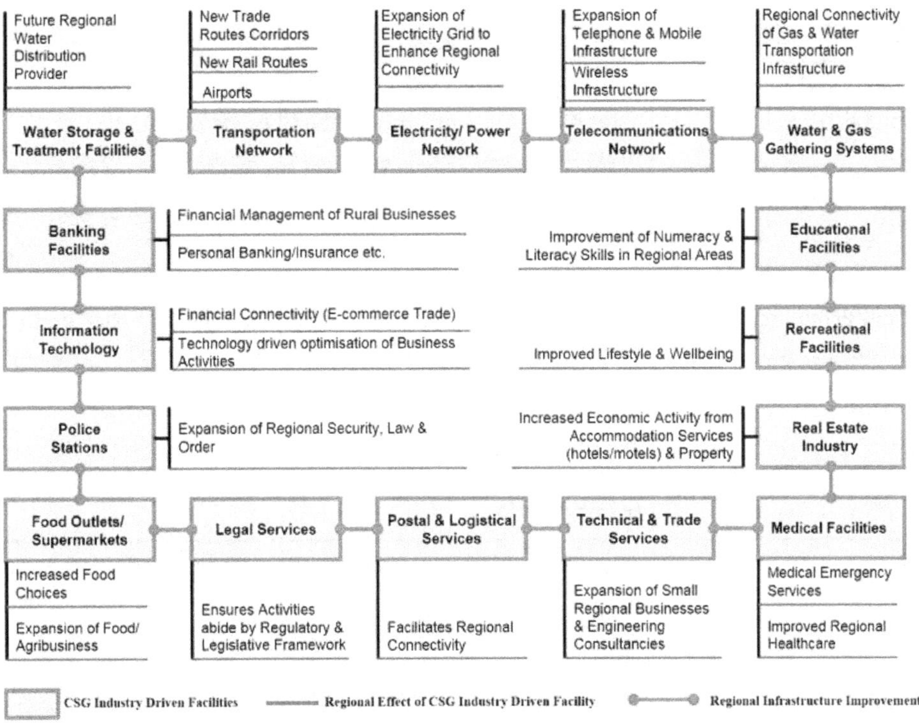

Figure 4. Effect of CSG industry driven facilities on regional infrastructure.

Mehreen &Underschultz [39], have investigated and analysed several industries that could potentially use the CSG by-products using screening matrices. Those industries with a link or contribution to the agricultural value chain, are natural candidates for implementation in the present agriculturally-rich CSG development areas. The screening matrix criteria are presented in Table 3. A comprehensive literature review was conducted by Mehreen &Underschultz [39] to assist the screening matrix analysis, whereby each criterion was rated (1 = low, 2 = medium, 3 = high). The scores were then totalled for each industry. Upon careful consideration and assessment of the applicability of each industry as a beneficial user of the CSG by-products, the screening matrix analysis revealed that industries that were closely associated with or contributing to the agricultural value chain (typically API's) scored highly. These API's with high coexistence potential with the CSG industry are: meat processing, irrigation, tanneries and livestock watering. An excerpt of the literature review with specific reference to these high-scoring API'S is summarised in Table 4. The agricultural landscape surrounding CSG developments is typically dominated by cattle grazing properties which are notably sustained by the API's that have scored highly in the screening matrix results. Therefore, the authors have placed a greater emphasis on analysing the cattle industry-based agricultural value chain for promoting coexistence opportunities with the CSG industry.

Table 3.Screening matrix criteria ([39]).

Screening matrix criteria	Description	Question guide
Environmental sustainability	Environmental impact from establishment of prescribed industry was considered as a vital criterion to assess its viability.	Is this option environmentally sustainable?/Does this option utilise a waste product of the CSG industry?
Location/Proximity (importance of location	The distance between the source of the CSG industry derived service and the end user for beneficial use was regarded as critical due to increased costs that may be associated with transportation.	Can the end user be in close proximity to the source location of the CSG industry derived service?
Reliability	There must be a consistent uptake of the CSG industry derived service by the proposed option for beneficial use for there to be an ongoing and 'reliable' coexistence of all industries. A point to consider is that there should be an adequate production of the service to meet high level demands from the end user, or alternatively, there must be a sufficient demand from the end user industry for a reliable uptake of the CSG industry derived service.	Will the end user regularly use the CSG industry derived service?

Screening matrix criteria	Description	Question guide
Technical feasibility	The potential co-existent industry should possess a high level of technological maturity for a high score in this criterion. Alternatively, industries with underlying technologies which are considered to be under research and development (R&D) phase were scored as having low technical feasibility.	Is the underlying technology mature and well known for the functioning/ establishment of the industry?
Community benefit	For a high score in this criterion, potential industries must directly inject benefit to the regional community near the CSG development. This benefit can be sourced from increased employment opportunities, increased social awareness of local businesses, and any facilitation of the regional community's wellbeing. Those industries that are regarded as having a justifiable negative impact from a social context have been considered as poor contributors to the advancement of the regional community.	Will the community benefit from this industry?
Social acceptance	For there to be co-existence of other industries alongside the CSG industry in the nearby regional area, there must be acceptance of receiving the CSG industry derived service from the regional community. Those options that are traditionally regarded as propagating community benefit from a social standpoint have been scored highly.	Will there be social acceptance for this industry? Are there any social repercussions associated with this industry?

Screening matrix criteria	Description	Question guide
Supporting workforce	Industries which require a workforce with skills that are already present in the CSG development area were considered as a great advantage, as it would promote the local employment sector without the need for upgrading skills or further training; consequently, these industries were scored highly.	Is there a supportive workforce already present in the regional area of interest for colocation/coexistence of this industry?

Table 4. Summary of literature for the beneficial use of CSGAW by high-scoring API's (modified from [39]).

Beneficial use option/Industry	Key criterion	References
Irrigation	Environmental sustainability	[4, 13, 47, 48, 49, 50]
	Location/Proximity (importance of location)	[4, 13]
	Reliability	[4, 13, 51]
	Technical feasibility	[13, 48, 52]
	Community benefit	[4, 51, 52]
	Social acceptance	[4, 51, 52]
	Supporting workforce	[53]
Livestock watering	Environmental sustainability	[4, 6, 54, 55]
	Location/Proximity (importance of location)	[13, 56]
	Reliability	[6, 13, 57, 58]
	Technical feasibility	[13, 59]
	Community benefit	[13, 59]
	Social acceptance	[6, 13]
Abattoir/Meat processing industry	Environmental sustainability	[60, 61]
	Location/Proximity (importance of location)	[13, 62, 63, 64, 65]
	Reliability	[60, 61, 63]
	Technical feasibility	[60, 61, 66, 67]
	Community benefit	[66]
	Social acceptance	[67, 68]
	Supporting workforce	[53, 69]

Beneficial use option/Industry	Key criterion	References
Tannery/Leather	Environmental sustainability	[70, 71, 72, 73, 74, 75]
	Location/Proximity (importance of location)	[13, 76, 77]
	Reliability	[4, 73, 77, 78, 79]
	Technical feasibility	[73, 76, 77, 78, 79]
	Community benefit	[73, 76, 77, 78, 79]
	Social acceptance	[76, 77, 78]
	Supporting workforce	[53]

3.1. Crop Cultivation–Irrigation

As Queensland's CSG operations are distributed across the agricultural landscape, the use of CSGAW for irrigation purposes, especially in its large production volumes is a practical option [48, 55]. A successful implementation of the CSGAW irrigation scheme is in the Australia Pacific LNG Project which is enabling the use of treated CSGAW for drip irrigation projects involving a 300 hectares (ha) Pongamia plantation (bio-fuel crop) [80].

Some extent of treatment or amendment of CSGAW is required prior to irrigation use [48]. Defective plant growth patterns have been exhibited by crops that have been experimentally irrigated with certain untreated CSGAW [4, 13]. The direct application of CSGAW for irrigation purposes is therefore impractical, in most cases as it quite often of a poorer quality than the present water source distributed for irrigation. The successful implementation of this water management option is highly dictated by the CSGAW quality parameters such as salinity and sodicity. As a basic minimum requirement, CSGAW must be treated for the removal of salt prior to use for irrigation purposes [48]. Irrigating low salt tolerant crops with raw CSGAW which is saline will cause crusting of soil structure, decreased water retention ability, and increased soil erosion from runoff, in turn defecting healthy crop growth [13]. Soil chemistry (salt levels, pH), climatic conditions, crop salt tolerance ranges, topography of land are also critical parameters that dictate the extent of water treatment required prior to irrigation use of CSGAW [4, 48]. Table 5 outlines the salt tolerance ranges for potential crop groups.

While CSGAW that has been treated in accordance with the regulatory standards [13, 81] may be argued as being safe to use for irrigation purposes, there has been some research that suggests that from a long-term perspective, there may be cumulative concentration of salts over time, which can pose a threat to soil structure. The impact of CSGAW irrigation and its associated environmental sustainability should be considered on a case-by-case basis as each site differs in its soil and crop profile [48]. The average water usage per property and subsequently irrigation extraction volume allocations should be implemented [4]. Providing CSG-sourced water for irrigation purposes would help in drought-proofing the land and improving land productivity (increased opportunity to harvest crop and livestock grazing yield), thereby boosting the economic potential for agribusiness and directly opening up potential investment opportunities such as food-based tourism [52].

Table 5. Crop suitability based on irrigation water salinity ([13, 39]).

TDS (mg/L)	Water salinity rank	Crop suitability	Potential crop
<390	Very low	High sensitivity	Flowers/fruits
390–780	Low	Reasonable sensitivity	
780–1740	Medium	Reasonable tolerance	Clover
1741–3120	High	Tolerant crops	Corn, Lucerne, sorghum, soy bean
3121–4860	Very high	Highly tolerant crops	Cotton, cereals (wheat), barley
>4861	Extreme	Usually too saline	—

3.2. Livestock Watering

Land areas that are dominated by grazing activities and animal farming require feedlots facilities for providing fodder and water to animals, prior to slaughter. Such feedlot facilities require an adequate supply of water for animal consumption (livestock watering). Using CSGAW for the feedlots industry assists in providing water supply to drought affected areas which can allow the functioning of the livestock industry which will directly benefit the meat processing agriculture value chain to have a supply of livestock for slaughter. The tolerance range of livestock to the consumption of untreated CSGAW varies (Tables 6 and 7). Typically, the quality of CSGAW is regarded as being within the acceptable limits for livestock watering purposes [4]. There have been some cases where excess fluoride levels in the water (non-CSG sourced) supplied for livestock consumption have caused poor dental health (e.g. fluorosis) in the affected animals [13]. If raw/ untreated CSGAW is deemed unsuitable for direct livestock consumption, then low level CSGAW treatment must be implemented to eliminate the water from high TDS and fluoride levels, prior to livestock consumption [13].

Table 6. Livestock tolerance range for drinking water ([4, 39, 81]).

Livestock	Total dissolved solids (mg/L)		
	No adverse effects on animals expected	Animals may have initial reluctance to drink or there may be some scouring, but stock should adapt without loss of production	Loss of production and a decline in animal condition and health would be expected. Stock may tolerate these levels for short periods if introduced gradually
Beef cattle	0–4000	4000–5000	5000–10,000
Dairy cattle	0–2400	2400–4000	4000–7000
Sheep	0–4000	4000–10,000	10,000–13,000*
Horses	0–4000	4000–6000	6000–7000
Pigs	0–4000	4000–6000	6000–8000
Poultry	0–2000	2000–3000	3000–4000
Sheep on lush green feed may tolerate up to 13,000 mg/L TDS without loss of condition or production.			

Table 7. Effect of TDS levels on livestock ([12, 39]).

TDS (mg/L)	Results
<1000	Highly tolerable for all livestock
1000–2999	High satisfactory tolerance for all livestock with some cases of diarrhoea reported for livestock consuming the water source for the first time.
3000–4999	Satisfactory for livestock with some diarrhoea reported for livestock consuming the water source for the first time.
5000–6999	Reasonably safe however should be avoided for pregnant and lactating animals
7000–10,000	High risk to young offspring however older livestock may still consume the water supply
>10,000	Unsuitable for all livestock

3.3. Meat Processing

There is a growing demand for high-quality meat produce both in the domestic Australian and international markets. This is representative of the economic revenue associated with meat production, where ~ $USD 1.2 billion was generated from the sale of Australian meat products [82]. The Department of Agriculture, Fisheries & Forestry (DAFF) has reported that there is a lack of adequate meat slaughtering and processing sites in Queensland, Australia [83]. New slaughter and meat processing facilities must be constructed to meet the demand from international markets for Australia's high-quality meats. Constructing abattoir and meat processing facilities in grazing corridors would reduce the transportation costs associated with transferring livestock from grazing fields to slaughter houses and inject economic revenue to the agricultural value chain. Furthermore, this colocation would also reduce transportation costs associated with transferring treated CSGAW for use at the abattoir site. Approximately 44% of Australia's total cattle numbers are present in Queensland [84]. Abattoir facilities are heavy users of water particularly during slaughtering and downstream meat processing stages [61, 63]. An environmental concern by many already existent abattoirs is the typically high nutrient load of the effluent water, which cannot be directly used as fertiliser. In such cases, amended CSGAW can be mixed with the effluent stream as a diluting agent, making it viable as a nutrient rich fluid to be applied on crops. The economic potential of the meat processing industry from both a local and international standpoint is vital to the growth of Australia's agribusiness sector.

3.4. Tannery: Leather Processing Industry

The waste brine generated from CSGAW treatment can be beneficially used in leather manufacturing processes. Saline rich feed water (e.g. brine) is required for curing the hides, particularly for antibacterial purposes, as well as for degreasing processes [85]. The tannery facility may be constructed at a proximal distance from the CSGAW distribution and abattoir sites, to optimise costs associated with the transportation of water and hides. The leather processing industry is a viable user of water, however flows (treated CSGAW and brine) will be directly related to the number of carcasses processed at the abattoir, which will in turn have consequences for the number of hides produced for leather manufacturing.

Providing CSGAW and CSG industry-sourced brine to the leather processing industry has massive potential to inject new economic opportunities for the local economy and creates avenues for international export if produced on a large scale. Purposefully co-locating tannery facilities with CSGAW distribution sites, has the added advantage of processing recycled tannery effluent waste through the same water treatment site. This suggested industry would promote the agricultural value chain and provide a potential coexistence opportunity for both the CSG industry and an API.

4. AGRIBUSINESS PROMOTING INDUSTRIES: COEXISTENCE POTENTIAL WITH COAL SEAM GAS

As the 'native' industry in CSG operating areas is the agricultural industry and associated agribusinesses, it is important to facilitate the growth and progress of those industries. The concept of a supply chain is services from one entity flow to another entity, through a medium that allows the flow of services to take place. In this way, services of one industry can pass their benefit to another industry, thereby contributing to a supply chain type model. Similarly, services provided by the CSG industry (such as by-product CSGAW) to local agribusinesses, can help to facilitate the agricultural value chain by enhancing food productivity, injecting investment opportunities, promoting agri-based tourism and trade prospects. Mehreen &Underschultz [39] propose an agri-based industrial coexistence model which promotes local synergies between the CSG industry and local agribusinesses. The model given in Figure 5 schematically represents the potential synergies between entities involved in the cattle value chain and the CSG industry, specifically focussing on CSGAW (and brine in the case of leather processing). This co-location of agri-based industries around the CSG developments allows the growth of the agriculture value chain, increased employment opportunities, regional infrastructure growth, and enhanced utility infrastructure [33, 46, 86].

The CSG water treatment and distribution facility can deliver CSGAW that has been amended (to the respective regulatory standards) for irrigation to nearby agricultural farms. Feedlot operations are provided with fodder or other feed crop that has been harvested by the agricultural farms in the area. These agricultural farms may even provide livestock (e.g. cattle) grazing lands. Untreated or amended CSGAW (treated in accordance with respective regulatory guidelines) (Tables 6 and 7) can be provided to feedlot operations for livestock consumption. The feedlots near abattoir / meat processing facility in the area, can provide livestock for slaughter. Treated CSGAW provided to the abattoir, can be utilised during sterilisation, evisceration, slaughtering and other meat processing stages. An anaerobic digester (AD) can treat the feedlot and abattoir effluent streams (high organic load dominated by biologically hazardous material) to produce biogas (methane) and highly concentrated nutrient load (potential fertiliser).

Prior to using the fertiliser on agricultural crops, this nutrient load from the AD must be diluted with treated CSGAW from the CSG water treatment facility. This fertiliser can be commercialised as a selling product or can be provided to agricultural farms and grazing areas to grow crops. The biogas produced from AD can be processed for abattoir's energy use (equipment) or provided to the CSG operator as a supplementary methane source. The CSG Water treatment facility can provide the saline-rich CSGAW for leather processing in the tannery facility and also provide local water treatment capacity for otherwise unusable wastewater from meat processing and tannery operations. Note that other local services (telecommunication and transportation infrastructure, and services in regional towns) that have developed as a result of CSG development will have longer term sustainability if they are also servicing an expanding co-located agribusiness chain.

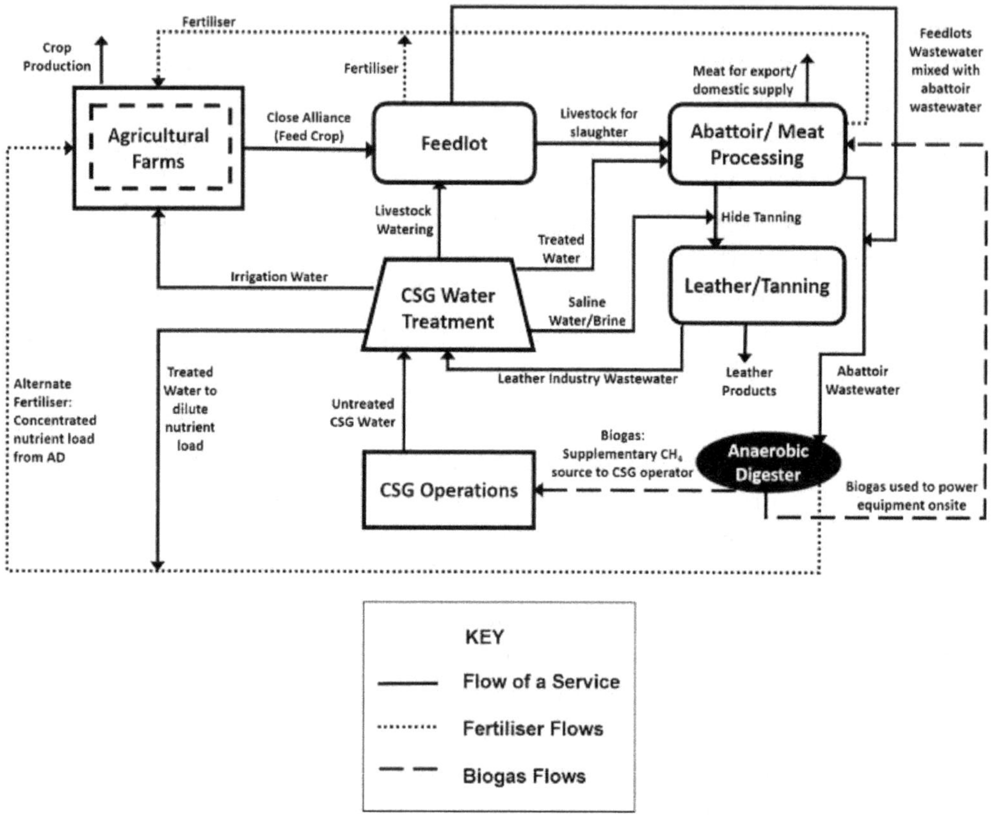

Figure 5. Suggested agri-based industrial coexistence model based on cattle value chain CH_4 = methane (modified from [39]).

The water requirements from each API in Figure 5 were calculated and compared with the modelled volumes of treated CSGAW for distribution from the CSG water treatment sites. This is summarisedin Figure 6. Some assumptions that were taken into consideration when calculating water consumption rates are as follows:

- The water required for irrigation (4300 kL/day) is calculated for 40 ha of agricultural land [87]
- The average cattle numbers processed at the abattoir are at a rate of 1400 cattle per day [88]
- The water consumption for processing one cattle hide in tannery operations is 702 litres (L) [70]
- Water consumption per cattle head at feedlot operations has been taken as 130 L/ cattle head [88, 89]
- Typical water treatment installed capacity of 20,000 kL/day which is taken as being available from the CSG water treatment facility [90]

Upon calculation of the water consumption in the entities involved within the agri-based co-existence model, it was noted that the demand (8406 kL/day) is much lower than the average water supply capacity. As the local labour workforce has an agri-based skillset, there would not be a skill shortage for the API's involved in this model. In fact this would help retain the local agri-based workforce with more job options. The main concern associated with the sustainability of this coexistence model may be the extent of water supply in the future as the CSGAW production volumes eventually fall. One option would be to use present piping and well injection infrastructure built for recharging aquifers, to collect and re-harvest the CSGAW for a sustainable supply of water into the future, when the CSGAW production has reached its end of life period.

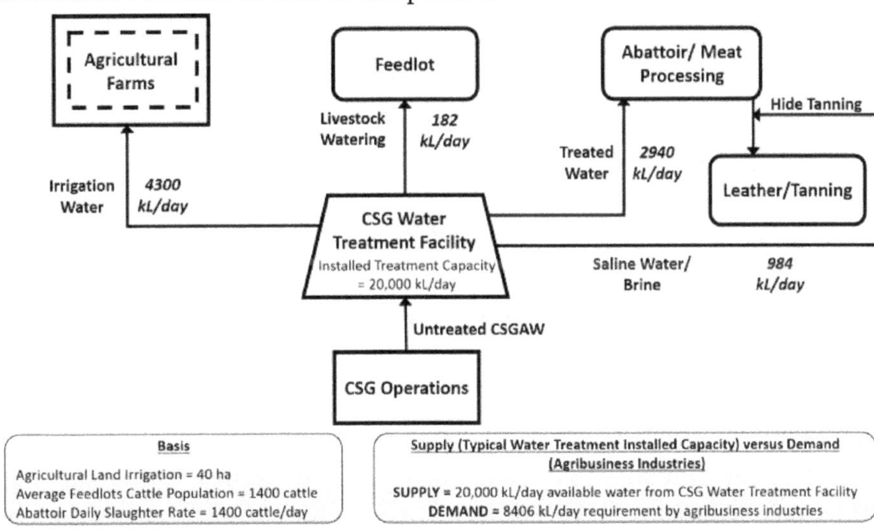

Figure 6. Typical water consumption by agri-business industries in comparison to coal seam gas water supply. kL = kiloliters (modified from [39]).

From an economic and community perspective, there is great value in promoting coexistence of agri-based industries alongside the CSG industry. However, the progress of amalgamating agricultural industries with the CSG industry has been slow [9]. There is cumulative effect of coexisting CSG developments in close proximity to agricultural developments that are complicated by community attitudes, local industries, environmental assets, and regulations [91, 92].

In regional CSG development, there is often concern for the preservation of environmental assets, particularly land and water as they provide economic value, ecological diversity, recreational value, and aesthetic value. As CSG developments are often located on prime agricultural land, land use conflict and stakeholder trust is a concern for gas operators [93]. A lack of trust in the CSG operator is quite often the most significant social issue which underpins many of the other concerns affecting the progress of promoting coexistence between agri-based industries and

CSG industry [91]. Land access agreements and their associated 'confidentiality clauses' can contribute to the distrust with CSG operators and regulatory bodies. Some government or CSG operator funded financial incentive is provided to landowners to promote greater cooperation [92]. Farmers with increased distrust in the CSG companies can have negative opinion of other farmers that have accepted monetary incentive. This can be viewed as having betrayed the 'rural fabric' that unites farmers and can create a local divide within the farming community. These social issues must also be addressed in order to better promote the coexistence value of the agribusinesses alongside the CSG industry. Strategic governance by federal and state governments to ensure trust with the local landowners must take effect to bridge the gap between agri-based industries and the CSG industry.

Analysing the effect of the CSG industry from a social perspective is quite often not as 'tangible' as analysing economic growth or environmental impact [92]. Perhaps this is attributed to the ability to better quantify economic and environmental impacts rather than social indicators which tend to be more of a qualitative nature. Therefore, conceptualising the potential impact on the social fabric underpinning the regional communities in the heart of CSG development regions can be difficult and may pose a barrier to better understand the effect of the CSG industry on the community from a social perspective. This further complicates analysis of the coexistence potential between the CSG industry and agri-based industries. It is therefore important to consider the cumulative impact of CSG development rather than the isolated impact.

When there are industries that are sharing infrastructure, there is an increased risk to the normal business case. For industries to coexist and gain mutual benefit, requires mutual trust. When the business risk is too high to share infrastructure between industries, it makes coexistence difficult. In this case, one company owns the infrastructure (e.g. CSG industry owned water treatment facilities) and another entity such as a new meat processing plant could benefit from utilising that business service. If access to water treatment is a critical component of the business case for the meat processing plant, but not in its control, this could pose an unacceptable risk to the establishment of the meat processing plant. For example agricultural wastes are characterised as having a high organic load, particularly in animal-based agri-based industries [61, 63]. Combining waste streams from such industries, and processing the produced wastewater through the CSG water treatment facility may increase the risk to the business model, and may pose as an unnecessary complication for the CSG operator. There must be corporative legislations that will be designed to remove the business risk; the support of the federal and state government is mandatory. Adequate planning must be implemented to remove such risks. The colocation reduces transportation costs dramatically due to the centralised location of the water treatment facility in relation to all the agri-based industries involved.

If agri-based industries are dependent on water, such a setup proposed in the agribusiness coexistence model in this paper, would mean that those industries will be heavily reliant on the CSG industry for providing water for their beneficial use. Due to the long period entailing the business case, it may be difficult to attract investment. This is perhaps another reason that has hindered the amalgamation of the CSG industry colocation with the agricultural industry sooner than later. Therefore, it will be important to find innovative business models that can alleviate these business risks and allow investment in a co-existence model where different industries can share infrastructure.

5. CONCISE SUMMARY

The agricultural industry can benefit from the by-products and services of the CSG industry due to their shared location and the similarity of required skillsets, minimizing the need for significant skills upgrades. This study has assessed the feasibility of utilizing CSGAW supply for irrigation, animal watering, meat processing, and leather manufacturing as recommended by the API. Water treatment is considered necessary before the API's may be used beneficially. By utilizing coal seam gas (CSG) by-products, especially CSG associated water (CSGAW), in conjunction with agricultural practices, it supports local agricultural industries and enhances the agricultural value chain in rural areas dominated by agriculture near CSG projects in Queensland. The agri-based industrial coexistence model enables agricultural processing industries to make use of by-products from the CSG industry for beneficial purposes, hence enhancing the sustainability and growth of the agricultural value chain. It offers the ability to serve as a "drought buffer" for landowners, helps sustain local skills, and creates long-term employment opportunities. Offering Controlled-Source Aquifer Water (CSGAW) for irrigating crops used for human and livestock consumption can stimulate the growth of the meat processing and leather manufacturing sectors, leading to increased land productivity and reinforcing the agricultural value chain. Additionally, placing the API's near the CSG water treatment facility would optimize the utilization of a centralized utilities and telecommunications infrastructure network. Reusing CSGAW through the existing infrastructure designed for regulated aquifer recharge is proposed as a way to recycle CSG-derived water for the API's once CSG production ends. Potential benefits of the agri-based coexistence model include higher employment rates, increased export trade, sustainable crop harvesting, and improved functioning of the local agricultural value chain. This model also has the potential to support the development of additional industries such as agri-tourism, biofuel production, and local meat and leather processing. The agri-based cohabitation model incorporates the agricultural value chain. It centralizes all agricultural industries involved, improving supply chain connectivity over short distances and significantly cutting transportation costs that would be incurred by moving raw products to other locations for further processing. Agricultural industry and agri-based product production typically originate from rural areas and regional towns that are linked by highways. This results in a fragmented value chain. The agribusiness coexistence model can help control the demand cycle for agricultural products by establishing a localized network of businesses involved in the agricultural value chain. The agribusiness coexistence model enables local consumers to buy fresh locally grown produce, made more available by irrigation water supplies, thereby aiding local farmers in sustaining the locally grown movement. The adoption of policies related to the agribusiness coexistence model can have a worldwide effect by exporting high-quality meats and other agricultural food products to international markets. This can bring in investments for Australia's economic growth and enhance the agricultural value chain. The proposed agricultural coexistence model has demonstrated the ability to simultaneously advance CSG operations alongside agriculture-based sectors, ensuring the sustainability of the energy-food relationship. Successful coexistence of 'API-CSG' necessitates meticulous planning and ongoing involvement with the local industry.

6. AT A GLANCE

Australia's coal seam gas (CSG development has not only contributed to government royalties but has also aided in regional economic growth, expansion of remote populations and facilities, increased job opportunities, and enhanced regional infrastructure, particularly in regional Queensland. Australia has significant economic opportunities by exporting the resource to global energy markets. Several present CSG installations in Australia are situated near key agricultural areas used for cattle grazing. Not recognizing possible chances for coexistence between agribusiness promoting industries (API's) and the CSG industry may hinder the agriculture value chain, thus limiting Australia's food security and agricultural export potential. The economic advantages of the CSG sector, along with the significance of a thriving agricultural sector, provide a basis for exploring prospects for coexistence between both businesses. The focus is on the possible synergies between the CSG business, particularly its by-products, and the local agricultural industry, which is usually dominated by API's.

REFERENCES

1. Teske S, Pregger T, Simon S, Naegler T, Graus W, Lins C. Energy [R] evolution 2010—A sustainable world energy outlook. Energy Efficiency. 2011;4(3):409-433

2. Girod B, van Vuuren DP, Hertwich EG. Climate policy through changing consumption choices: Options and obstacles for reducing greenhouse gas emissions. Global Environmental Change. 2014;25:5-15
3. Abousnina RM, Nghiem LD, Bundschuh J. Comparison between oily and coal seam gas produced water with respect to quantity, characteristics and treatment technologies: A review. Desalination and Water Treatment. 2014:1-16
4. Khan S, Kordek G. Coal Seam Gas: Produced Water and Solids. Report prepared for the Office of the NSW Chief Scientist and Engineer (OCSE). Sydney: The University of New South Wales. 2014
5. Hamawand I, Yusaf T, Hamawand SG. Coal seam gas and associated water: A review paper. Renewable and Sustainable Energy Reviews. 2013;22:550-560
6. Nghiem LD, Ren T, Aziz N, Porter I, Regmi G. Treatment of coal seam gas produced water for beneficial use in Australia: A review of best practices. Desalination and Water Treatment. 2011;32(1-3):316-323
7. Smith G, Waguih A, Redrup JP, Nick Kholgh A, Aguirre E, editors. Case study: Factory drilling Australia's unconventional coal seam gas. SPE Asia Pacific Oil & Gas Conference and Exhibition; 2014: Society of Petroleum Engineers
8. O'Kane M. NSW Chief Scientist and Engineer, Initial Report on the Independent Review of Coal Seam Gas Activities in NSW, July 2013. 2013
9. Towler B, Firouzi M, Underschultz J, Rifkin W, Garnett A, Schultz H, et al. An overview of the coal seam gas developments in Queensland. Journal of Natural Gas Science and Engineering. 2016;31:249-271
10. Jakubowski R, Haws N, Ellerbroek D, Murtagh J, Macfarlane D. Development of a management tool to support the beneficial use of treated coal seam gas water for irrigation in eastern Australia. Mine Water and the Environment. 2014;33(2):133-145
11. Tan P-L, George D, Comino M. Cumulative risk management, coal seam gas, sustainable water, and agriculture in Australia. International Journal of Water Resources Development. 2015;31(4):682-700
12. ALL Consulting. Handbook on coal bed methane produced water: Management and beneficial use alternatives. Tulsa, OK, USA. 2003
13. DNRM (Department of Natural Resources and Mines). Coal Seam Gas (CSG) Water Management Study. Brisbane, QLD, Australia: Department of Natural Resources and Mines; 2004
14. Underschultz J. Unconventional Gas. 2016

15. Guo P, Cheng Y. Permeability prediction in deep coal seam: A case study on the no. 3 coal seam of the southern Qinshui Basin in China. The Scientific World Journal. 2013;2013:1-10
16. Papendick SL, Downs KR, Vo KD, Hamilton SK, Dawson GK, Golding SD, et al. Biogenic methane potential for Surat Basin, Queensland coal seams. International Journal of Coal Geology. 2011;88(2):123-134
17. Speight JG. An Introduction to Petroleum Technology, Economics, and Politics. Hoboken: John Wiley & Sons; 2011
18. Duus S. Coal contestations: Learning from a long, broad view. Rural Society. 2013;22(2):96-110
19. Bao Y, Wei C, Wang C, Li L, Sun Y. Geochemical characteristics and identification of thermogenic CBM generated during the low and middle coalification stages. Geochemical Journal. 2013;47(4):451-458
20. Fallgren PH, Jin S, Zeng C, Ren Z, Lu A, Colberg PJ. Comparison of coal rank for enhanced biogenic natural gas production. International Journal of Coal Geology. 2013;115:92-96
21. O'Keefe JM, Bechtel A, Christanis K, Dai S, DiMichele WA, Eble CF, et al. On the fundamental difference between coal rank and coal type. International Journal of Coal Geology. 2013;118:58-87
22. Xu CW, editor. Exploration Prospect of low rank Coalbed methane coal in Huhehu depression Hailar Basin. Advanced Materials Research; 2013: Trans Tech Publ
23. Ward C, Kelly B. Background Paper on New South Wales Geology: With a Focus on Basins Containing Coal Seam Gas Resources. UNSW Global Pty Limited. 2013
24. Close JC. Natural Fractures in Coal. Chapter 5: Hydrocarbons from Coal. Resource Enterprises Inc. Salt Lake City, Utah, USA. 1993
25. Kresse TM, Warner NR, Hays PD, Down A, Vengosh A, Jackson RB. Shallow groundwater quality and geochemistry in the Fayetteville Shale gas-production area, north-central Arkansas, 2011. US Geological Survey; 2012. Report No.: 2328-0328
26. Schoell M. The hydrogen and carbon isotopic composition of methane from natural gases of various origins. Geochimica et Cosmochimica Acta. 1980;44(5):649-661
27. George JS, Barakat M. The change in effective stress associated with shrinkage from gas desorption in coal. International Journal of Coal Geology. 2001;45(2):105-113

28. Cook P. Life Cycle of Coal Seam Gas Projects: Technologies and Potential Impacts. Report commissioned for the independent review of coal seam gas activities in NSW by the NSW Chief Scientist: PJC International Pty Ltd. 2013

29. Shaw M. Stream Ecosystem Health Response to Coal Seam Gas Water Release: Hazard Characterisation. Department of Science. Information Technology, Innovation and the Arts: Brisbane; 2010

30. Williams R, Walton A. Community Expectations and Coal Seam Gas Development: A report to the Gas Industry Social and Environmental Research Alliance (GISERA). January 2014. CSIRO report, CSIRO, Canberra. Available at GISERA website; 2014

31. APPEA (Australian Petroleum Production and Exploration). Australian LNG set to ride out low price storm report 2016 [Available from: https://aogexpo.com.au/lng/australian-lng-set-to-rise-out-low-price-storm-report/

32. Dunlop J, McGregor G, Rogers S. Cumulative Impacts of Coal Seam Gas Water Discharges to Surface Streams in the Queensland Murray–Darling Basin: Assessment of Water Quality Impacts. Brisbane, Australia: Department of Information Technology, Innovation and the Arts; 2013

33. Huth NI, Cocks B, Dalgliesh N, Poulton PL, Marinoni O, Garcia JN. Farmers' perceptions of coexistence between agriculture and a large scale coal seam gas development. Agriculture and Human Values. 2014:1-17

34. Davies P, Gore D. Background Paper on Produced Water Solids in Relation to Coal Seam Gas Production. Sydney, Australia: Macquarie University; 2013

35. Flukes P. GLNG Gas Field Development - Associated Water Discharge Study. Victoria, Australia: South Bank; 2009

36. AGL (Australia Gas Light Company). Upstream Investments Pty Ltd Camden Gas Project. Quarterly produced water quality monitoring report. North Sydney, NSW, Australia; 2013

37. Stearman W, Taulis M, Smith J, Corkeron M. Assessment of geogenic contaminants in water co-produced with coal seam gas extraction in Queensland, Australia: Implications for human health risk. Geosciences. 2014;4(3):219-239

38. Jangbarwala J. CBM-produced water. Water Conditioning & Purification. 2007

39. Mehreen SU, Underschultz JR. Coexistence opportunities for coal seam gas and agribusiness. Journal of Industrial Ecology. 2016

40. Tan P-L, Baldwin C, White I, Burry K. Water planning in the Condamine alluvium, Queensland: Sharing information and eliciting views in a context of overallocation. Journal of Hydrology. 2012;474:38-46

41. Duong HC, Chivas AR, Nelemans B, Duke M, Gray S, Cath TY, et al. Treatment of RO brine from CSG produced water by spiral-wound air gap membrane distillation—A pilot study. Desalination. 2015;366:121-129

42. McCusker K, Quinlan L, editors. Coal Seam Gas Water Treatment Facilities In Australia. Offshore Mediterranean Conference and Exhibition; 2013: Offshore Mediterranean Conference

43. QGC (Queensland Gas Company). Ensuring Responsible CSG Water Management and Beneficial Use Report. Brisbane, Queensland, Australia; 2010

44. Kempener S. Exploring strategies for water and salt management. Queensland Gas Conference; Brisbane, Australia. 2011

45. Walton AM, McCrea R, Leonard R, Williams R. Resilience in a changing community landscape of coal seam gas: Chinchilla in southern Queensland. Journal of Economic and Social Policy. 2013;15(3):0_1

46. Measham TG, Fleming DA. Impacts of unconventional gas development on rural community decline. Journal of Rural Studies. 2014;36:376-385

47. APLNG (Australia Pacific LNG Pty Limited). Environment & water management Report. 2013

48. Biggs AJ, Witheyman S, Williams K, Cupples N, De Voil C, Power R, et al. Assessing the salinity impacts of coal seam gas water on landscapes and surface streams: Department of Natural Resources; 2012

49. Brannock M, Page S, Broom G, Griffiths D, Rubinstein I. The Water Supply Spectrum: A Comparison of New Water Sources. OzWater Conference; 8-10 March; Brisbane, Australia. 2010

50. DNRM (Department of Natural Resources and Mines). Assessment of alternative use options for coal seam gas water proposed for Central Condamine Alluvium recharge schemes. 2013

51. Greet N, editor Changing Australia's view of energy security will provide for interconnected energy policy and a sustainable future. Dynamic ecolibrium: Sustainable Engineering Society Conference (SENG 2015); 2015: Engineers Australia

52. APLNG (Australia Pacific LNG Pty Limited). Combabula coal seam gas water management plan. Brisbane, QLD, Australia: Australia Pacific LNG Pty Limited; 2010

53. QGSO (Queensland Government Statistician's Office). Queensland regional profiles database. Queensland Treasure and Trade; 2014

54. DERM (Department of Environment and Heritage Protection). Approval of coal seam gas water for beneficial use: Environmental Protection (Waste Management) Regulation 2000; 2000

55. Grinter J. Bowen EIS - Surface water technical report prepared for Arrow Energy Pty. Ltd. Brisbane, Australia; 2012

56. Navi M, Skelly C, Taulis M, Nasiri S. Coal seam gas water: Potential hazards and exposure pathways in Queensland. International Journal of Environmental Health Research. 2015;25(2):162-183

57. DNRM (Department of Natural Resources and Mines). Healthy Headwaters Coal Seam Gas Water Feasibility Study. Business Case: Assessment of options for using CSG water in the Central Condamine Alluvium. 2013

58. Millar GJ, Couperthwaite SJ, Moodliar CD. Strategies for the management and treatment of coal seam gas associated water. Renewable and Sustainable Energy Reviews. 2016;57:669-691

59. Perez A, Middlemis H, Johnson K, Riches V. Reuse and Disposal Options for CSG Produced Water: The Gloucester Gas Project, NSW. 2014

60. AMPC (Australian Meat Processor Corporation). Australian Meat Processor Corporation Report: Wastewater Environmental Best Practice Manual. 2012

61. MLA (Meat & Livestock Australia). Water consumption: Meat & livestock Australia Report. North Sydney, Australia; 2014

62. EPAWA (Environmental Protection Authority Western Australia). Report and Recommendation of the EPA: Proposed Abattoir Aurisch Road GinginD'Orsogna Bros Pty Ltd; 1990

63. Johnson S. Proposed updating and reopening of the Toodyay Abattoir at Lot 590 Church Gully Road, Toodyay: Report and recommendation of the Environmental Protection Authority. Perth, Australia; 1990

64. MRC (Maranoa Regional Council). Agricultural Competiveness Issue Paper. 2014

65. NTG (Northern Territory Government). Meat Processing Facility Project: Assessment Report 68 by the Australian Agricultural Company for the Environment and Heritage Division Department of Natural Resources, Environment the Arts and Sport; 2012

66. DAFF (Department of Agriculture FaF. Northern Australian beef industry: assessment of risks and opportunities. 2012

67. RDA (Regional Development Australia). Growing central Queensland Review. Fitzroy and Central West, Australia; 2015

68. DTPLI (Department of Transport PaLI. Local Planning Policies. Hepburn Planning Scheme: Department of Transport, Planning and Local Infrastructure. State Government of Victoria. 2013
69. AGPS (Australian Government Publishing Service). Industry Commission Report: Meat Processing. 1994
70. Bosnic M, Buljan J, Daniels R. Pollutants in Tannery Effluents: Report Prepared by United Nations Industry Development Organization for Pollution Control in the Tanning Industry in South-East Asia. Vienna, Austria; 2000
71. Buljan J, UNIDO (United Nations Industrial Development Organization). Costs of Tannery Waste Treatment. 15th Session of Leather and Leather Products Industry Panel; 10-14 September; León, France. 2005
72. Buljan J, Kral I. Introduction to Treatment of Tannery Effluents. Report prepared for the United Nations Industry Development Organization. Vienna UNIDO Leather Panel. 2011
73. Money C. Tannery waste minimization. The Journal of the American Leather Chemists Association (USA). 1991
74. Paul H, Antunes APM, Covington AD, Evans P, Phillips PS. Bangladeshi leather industry: An overview of recent sustainable developments. Journal of the Society of Leather Technologists and Chemists. 2013;97(1):25-32
75. Song Z, Williams C, Edyvean R. Treatment of tannery wastewater by chemical coagulation. Desalination. 2004;164(3):249-259
76. Fisher H, Pearce D. Salinity reduction in tannery effluents in India and Australia Centre for International economics. Impact Assessment Series Report 61. 2009
77. UNIDO (United Nations Industrial Development Organization). Future Trends in the world leather and leather products industry and trade. 2010
78. Money CA. Addressing salinity in tannery effluents and uses for land irrigation in Australia and India. World Leather. 2005;18(1):22
79. Money C, Ramasam T, Babu C, Muralidharan J, Ragava Rao J, Saravanan P, et al. Salinity reduction in tannery effluents in India and Australia. Report prepared for Australian Centre for International Agricultural Research (ACAIR). 2009
80. Moser A. Coal seam gas water management River Symposium 21-23 September; Brisbane, Australia
81. ANZECC (Australian and New Zealand Environment and Conservation Council). Australian and New Zealand guidelines for fresh and marine water quality. Vol 1 The guidelines (chapters 1-7). 2000

82. AIG (Australian Industry Group). Meat & meat product manufacturing 2013 [Available from: http://pdf.aigroup.asn.au/environment/7082_WPA_fact_sheet_MEAT.pdf

83. Gleeson T, Martin P, Mifsud C. Northern Australian Beef Industry: Assessment of Risks and Opportunities: Australian Bureau of Agricultural and Resource Economics and Sciences (ABARES); 2012

84. ABS (Australian Bureau of Statistics). Agricultural commodities, national and state—2011-12 data cube. 2012

85. Rydin S, Michael B, Scalet B, Canova M. Tanning of Hides and Skins. Best available techniques reference document. Seville, Spain 2013

86. Letts L. Coal seam gas production–friend or foe of Queensland's water resources? Environmental and Planning Law Journal. 2012;29(2):101-112

87. AGL (Australia Gas Light Company). Soil quality monitoring and management: Report 1—Pre irrigation. Sydney, NSW, Australia; 2010

88. Johns M. Water Collection and Data Analysis. Report prepared for Meat & Livestock Australia. North Sydney, Australia; 2011

89. Bonner S, Davis R, Gernjak W, O'Keefe M, Poad G, Scobie M, et al. Treatment Technologies for Feedlot Effluent Reuse. Report prepared for Meat & Livestock Australia. North Sydney, Australia; 2011

90. GWI (Global Water Intelligence). Sizing up the Coal Seam Gas Water Market. Global Water Intelligence. 2012

91. Gillespie N, Bond CJ, Downs V, Staggs J. Stakeholder trust in the Queensland CSG industry. The APPEA Journal. 2016;56(1):239-246

92. Collins N, Everingham J-A, Baumgartl T, Vink S, Rodriguez D, Cavaye J, et al. Energy resources from the food bowl: an uneasy co-existence. Identifying and managing cumulative impacts of mining and agriculture. 2013

93. Baker G, Slater S, editors. The increasing significance of coal seam gas in eastern Australia. PESA Eastern Australasian Basins Symposium III; 2008 14-17 September; Sydney

12 ENHANCING PRODUCTIVITY THROUGH DEVELOPING SUPPLIER PRODUCTION CAPABILITIES IN AGRICULTURAL VALUE CHAINS

1. INTRODUCTION: AN OVERVIEW

Understanding the scattered value creation processes that occur across different locations and time is essential in the context of global economic activity, leading to an interest in global value chains (GVCs). Engagement in the global value chain is considered essential for the continued existence and longevity of a company. Development partners have elevated GVC involvement to a prominent position on the policy agenda as a recommended strategy for enhancing agribusiness productivity and competitiveness in developing nations, particularly in sub-Saharan Africa. Development organizations are enhancing their organizational competencies to engage in Global Value Chains (GVC).[1] The World Trade Organization emphasizes the significance of participating in Global Value Chains (GVCs) by stating that any international trade and investment policy discussion that does not recognize the important role of GVCs would be outdated and of doubtful relevance [1].

Despite the interest in GVC participation, intense debate still lingers over root causes of why some countries are advancing in the global marketplace, while others are failing to do so [2]. Supply chain production capabilities to drive productivity gains for growth have been highlighted by the value chain (VC) fraternity researchers as an area of interest to investigate this phenomenon. In Adam Smith's classical work, entitled The Wealth of Nations [3], Adam contended that productivity gains are vital to the economy, and it is the true measure of a nations growth, as more (higher output) is being accomplished with less (minimum inputs). Capital and labor are both scarce resources, so maximizing their impact is a core concern of modern business. Productivity gains have been identified to emanate from technological advancements such as information and communication technologies, supply chain and logistics improvements, and improved skill levels within the workforce.

This study investigated the presence of productivity gains among primary producers, that is, the supply workforce (farmers or primary commodity producers) capabilities in Uganda's agribusiness value chains. The level of available supply production capabilities determines the choice of the governance structure to coordinate interfirm relationships, which can be done either through spot markets (arm's length transactions and/or cash transactions with immediate delivery) or hierarchies (production of goods and services by single integrated firm) [4]. Ref. [2] GVC framework, considered to be an extension of Williamson's transaction cost economics (TCE) theory, offers intermediary networks or quasi-hierarchies [5] for interfirm relationship coordination and production organization in the form of modular, relational, and captive value chains. The existence of the various networks for the organization of production and coordination of interfirm relationships implies the existence of both vertical and horizontal linkage mechanisms.

Porter [6] argued that the existence of comparative advantage, economies of scale, and excessive vertical integration are no longer sources of competitiveness and innovation. He contended that close linkages between buyers, suppliers, and other institutions are now the source for firm, industry, and sector competitiveness, as reflected in productivity [7]. This statement affirms that the level of strength of collaborative relationships in both vertical and horizontal linkages is the source for productivity gains.

According to Navdi and Halder [8], the cluster theoretical approach assists in assessing the gains of clustering (i.e., grouping similar objects) as a result of joint action, while the VC approach explores vertical linkages between firms and external actors. Although both approaches offer complementary synergies to each other, they do not elaborate on the measures applicable in assessing the strength of the interfirm relationships [2, 9]. This missing link is filled by the transaction cost economic approach [4], which analyzes investment transactional costs involved in interfirm relationships. Williamson [4] identified the transactional costs in the form of specific investment asset specificity, uncertainty, frequency of transactions, and opportunism. In this study, we contend that building supplier production capabilities involves undertaking investment asset specificity in terms of provision of inputs, knowledge, and skills transferred to primary producers with a purpose of building production capabilities for productivity gains. It also involves transactional costs related to searching, monitoring, and enforcement of contracts by either party, costs considered as eroding profit margins. More investments in building mutual trust relationships are crucial in order to minimize opportunism. This study adopted these measures in assessing the strength of collaborative vertical and horizontal linkages for the building of supplier production capabilities.

This research aimed to enhance the comprehension of the relationship between supplier/growers' production capacities and productivity growth in the Global Value Chain (GVC) of sugarcane and forestry sectors, providing valuable insights into the evolving GVC theory. This was accomplished by examining the effectiveness of value chain activities in vertical and horizontal connections between growers and buyers/millers, as well as among growers. This study compared supplier production capacities in the commercial sugarcane and forestry sectors in Uganda to determine their impact on industry productivity gains. The inquiry was guided by the following research questions:

- How does investment asset specificity in vertical and horizontal linkages explain performance differences between commercial sugarcane and forestry sector value chains?
- How does coordination of transactional costs in vertical and horizontal linkages explain performance differences between commercial sugarcane and forestry sector value chains?
- How do actors' behaviors in vertical and horizontal linkages explain performance differences between commercial sugarcane and forestry sector value chains?

2. METHODOLOGY

This research employed a case study approach as the major research strategy, with a survey complementing the results of the case study. The purpose of this research was to contribute to emerging GVC theory [2] and hence suitability of case study [10].

Field-based and multiple data collection methods (questionnaire survey, interviews, archives, and observations) were used to gather empirical data to address the research questions and for triangulation of results [10, 11, 12, 13, 14].

Industry-specific information sources such as Uganda Sugar Cane Technologist Association (USCTA), Uganda Sugar Manufacturers Association (USMA), and Sawlog Production Grant Scheme (SPGS) project provided expected industry productivity baselines [15, 16, 17]. The total number of registered producers was 389 of which 298 where functional as per SPGS performance report 2012/2013. Kinyara Sugar Works Limited provided growers productivity data for the years 2010/2011 to 2012/2013 financial years. The total number of out-growers in Kinyara Sugarcane Cluster was approx. 6000, of which 105 were registered and approximately 77 commercial producers were considered functional. The total number of anticipated respondents participating in the study was 100 (survey tool) and 20 (qualitative tool or interviews), from both forestry and sugarcane value chains. The response rate was 46 to the questionnaires and 9 to the interviews in forestry industry and 32 to the questionnaires and 10 to the interviews, respectively. The overall response rate was 81%. The sample size of 97 was generally found appropriate for studies of this nature, in line with [10] recommendation of cases between 4 and 10 as appropriate.

The cases involved in the entire value chain as the principal unit of analysis are explored and analyzed at three sublevels: micro (primary producers/growers' enterprises), meso (industry experts, miller's representatives, and association executives in the value chain), and macro (assessment of national policies and regulations). Principal component analysis was run for purposes of grouping items. Empirical data was analyzed using within case analysis that enabled intimate familiarity with each case as stand-alone entity and cross-case pattern analysis that enabled constant comparison of theory and data—iterating toward a fit between theory and data. Figure 1 represents a summary of the methodology procedure.

2.1. Measures of Productivity Gains

Adam Smith [3] in his book of the wealth of nations identified three productivity measures, namely, farm output, manufactured goods, and labor to produce goods. This study was interested in productivity gains in agribusiness and therefore considered farm output as the leading measure for farm enterprise productivity. The study adopted country-specific industry reports published by the Uganda Sugar Manufacturers Association (USMA) which consider farm output of 100 tons/ha as the baseline productivity measure of sugarcane maturity of 18–20 months [17]. The forestry sector productivity reports were obtained from the Sawlog Production Grant Scheme, assessing performance of growers and providing indicative desired contract performance measures of 90% in agronomical practices.

2.2. Measures of Strength in Vertical and Horizontal Collaboration for Building Supplier Production Capabilities

Vertical links in the value chain discipline serve as channels for the exchange of learning, skills, information, technical, financial, and business services amongst firms within the value chain. Horizontal links denote extended collaborative agreements between enterprises that entail interdependence, trust, and resource sharing to achieve collective action or common objectives.

The transactional cost approach provides measures for assessing the strength of the market coordination mechanisms [4]. The measures consist of investment asset specificity costs (knowledge, skills, and physical production inputs), coordination transaction costs (timely payments, information search, contract monitoring, and enforcement), transaction frequency, and relationship quality (mutually beneficial or exploitative). This study evaluated vertical and horizontal collaboration in enhancing supplier production capabilities within value chains, excluding the frequency of transactions as a measure.

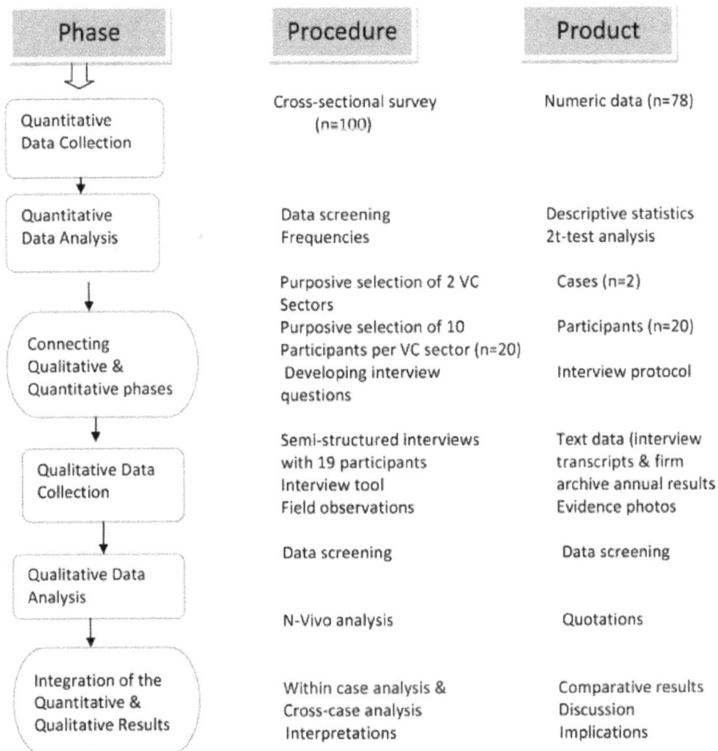

Figure 1. Research methodology flow chart.

3. RESULTS AND DISCUSSION

Research question 1: How does investment asset specificity in vertical and horizontal linkages explain performance differences between commercial sugarcane and forestry sector value chains?

The findings of the study suggested strong collaborative vertical linkages for building supplier production capabilities, as can be attested with the two-sample t-test. A comparison of the mean value for asset specificity in vertical relationships revealed that there was a significant difference in the mean rating by sector (forestry-sugarcane), Pr $(T < t) = 0.0002$. This means that there was a stronger collaborative relationship between millers and producers for inputs support, knowledge, and skills transfer in the sugarcane sector compared to the forestry sector. However, a nuanced view by qualitative data revealed that transfer of knowledge, skills, and inputs support was through contractors and not directly to primary sugarcane producers as evidenced by the quotations below:

"Planning is one of the biggest problems. The basic thing which a farmer gives is land... since farmers get inputs and using contractors for land preparation, harvesting and transporting, this creates sluggishness. i.e. some farmers think cane is for Kinyara, but now we are telling them to take it as a business so that they can plan, and save. Further, many farmers have no records...we hope that if farmers take up operations, they will develop capacity and protect their investments" (Out-Grower Manager, Sugar Mill).

The out-grower manager's statement was also validated by a primary sugarcane producer as stated below:

"Under the previous miller, harvesting, loading and transporting were done by the farmers' company - which was the business arm of the association. When current management came on board, this was abolished and they preferred to use contractors. With the previous miller knowledge was gained; we used to have courses in Kampala which was helpful, however, with the current miller not much has been gained" (Respondent Sugar).

Therefore, the qualitative findings above suggest weaker collaborative relationships for building production capabilities between millers and growers in the sugarcane sector. Further, the quotation below corroborated the quantitative findings of weak collaborative relationships for building production capabilities between millers and growers in the forestry sector.

"No miller is supporting growers, Nile Ply just buys from growers but without supporting them" (Program Manager, SPGS/Forestry).

On the other hand, the forestry sector had strong collaborative horizontal linkages existing with respect to investment asset specificity. A comparison of the mean value for asset investment specificity in horizontal relationships revealed that there was a significant difference in the mean rating by sector (forestry-sugarcane), Pr $(T > t) = 0.0023$. This means that there was a stronger collaborative relationship among producers and/or producer support agencies for inputs support, knowledge, and skills transfer in the forestry sector than the sugarcane sector. The above quantitative data finding was supported by qualitative data findings below:

"SPGS support has enabled at least 30% to improve production planning skills.... The change is more than significant. availability of forest valuation guidelines also gives growers basics on what price they cannot go below (reserve price) during negotiations in order to realize a return on their investments" (Program Manager—SPGS/Forestry).

The quote above suggests that transfer of knowledge, skills, and inputs support occurred directly to forestry producers from development agencies. This was evidenced by the t-test results that showed very strong significant differences (forestry-sugarcane), Pr $(T > t) = 0.0000$, regarding access to both technical and financial support from farmers' development agencies. This finding is also corroborated by both foreign and local tours (see Figures 2 and 3) organized by SPGS—a development support agency in conjunction with UTGA—the National Farmers Association, whose aim is to equip forestry growers with technical knowledge for building production capabilities.

The findings render support for strong horizontal collaborative relationships for building production capabilities. Findings also complemented with the evidenced adduced in the quotation below:

"Growers have acquired technical competency in plantation establishment, maintenance such as thinning, pruning, marking and harvesting....Good relationships exist, especially those under UTGA. When we call cluster meetings, we see the will to share, cooperate, and avail their plantations for study" (Association General Manager, UTGA/Forestry).

Figure 2. Ugandan forestry commercial farmers learning and nurturing of quality tree seedlings. Source: Primary field data courtesy of SPGS study tour Mondi nursery facility (South Africa).

Figure 3. Ugandan forestry commercial farmers on networking and information sharing testing the strength of the pole required by the market. Source: Primary field data courtesy of UTGA/SPGS local study tour New Forestry Company Ltd—pole treatment plant in Uganda.

The statement by one of the growers below did not discount the above statements:

"Yes, knowledge through newsletters, client meetings... You access information on prices, even if someone (buyer/miller) comes with a monopoly, but he realises that you are able to chip in from an informed position" (Respondent, Forestry).

However, with respect to sugarcane sector, qualitative data further supported quantitative data that there were no gains in building production capabilities from horizontal collaborative linkages as evidenced with the quotation below:

"Percentage wise in knowledge and skills transfer is still low, ... the previous association (KSGL) incurred liabilities, hence farmers lost the trust but now picking up slowly" (Association Chairman, Sugarcane Sector).

Our observations on the question about the influence of investment asset specificity in vertical and horizontal linkages on performance differences between commercial sugarcane and forestry sector value chains is as follows. The sugarcane sector exhibited high investment asset specificity in vertical linkages between primary producers and millers. This was evidenced by quantitative data which suggested strong collaborative vertical linkages for building supplier production capabilities. However, a nuanced view of the qualitative data pointed to the existence of production capabilities in the vertical linkages but residing in the use of contractors rather than the cane growers themselves. This finding was supported by a study of small-scale growers in the South African sugar industry that arrived at similar results [18]. Therefore, this finding suggested that growers were heavily dependent on the millers and contractors employed by the millers to offer services to the contracted growers. It was this level of high dependency of growers upon millers that could offer plausible explanations for failure to develop production capabilities among most growers in the Ugandan sugarcane sector.

On the other hand, quantitative data revealed a stronger collaborative relationship among producers and/or producer support agencies for availing inputs, knowledge, and skills transfer in the forestry sector than the sugarcane sector. This finding was corroborated by qualitative data and validated by observatory field data, which confirmed that the transfer of knowledge, skills, and inputs support occurred directly to forestry primary producers. This was done by development partner agencies through provision of both technical and financial support accompanied with foreign and local exposure learning platforms (see Figures 2 and 3), with a purpose of building production capabilities for achieving productivity gains. This finding supports the GVC literature, as argued by GVC proponents such as [19, 20] that a combination of technical and investment support in highly governed chains explains how relatively underdeveloped regions become major export producers in a short period of time. They cited the example of the Brazilian shoe industry in the 1970s and the Vietnamese garment industry in the late 1990s. Made a similar observation in his study of the textile and garment supply chain in South Africa [14]. He found out that companies that had closer collaboration in training and assistance attained a higher diffusion of skills in a shorter time to achieve supply chain efficiency levels required to compete effectively. This scenario can be described as a true reflection of Uganda's evolving commercial forestry sector, which enjoys support from development agencies and producer associations that links both primary producers and millers.

Further, the proponents of the cluster theoretical framework argue that diffusion of production capabilities is not only limited to GVC participants, but there is also knowledge and skills "spillover" in a geographical area and/or localities of business operations [8, 21, 22, 23, 24, 25, 26]. They argued that impact of knowledge and skills "spillover" accounts for the rise of entrepreneurship in various forms such as functional upgrading, new entrants in the existing clusters and value chains, and the start of new parallel competitive value chains. This entrepreneurial potential can be said to be available especially in Uganda's commercial forestry value chain sector, only if other agro-commodities can have a ready market for commercial production.

The findings above explained the investment asset specificity in vertical and horizontal linkages for interfirm relationships. However, as firms engage in the exchange process, they may be vulnerable to coordination transactional costs and opportunism by either party involved in the execution of the contracts. The next section presents and discusses comparative results with respect to vulnerability to transactional costs in both vertical and horizontal collaborative relationships for building production capabilities in order to achieve productivity gains.

Research question 2: How does coordination of transactional costs in vertical and horizontal linkages explain performance differences between commercial sugarcane and forestry sector value chains?

A comparison of the mean value for transactional costs in vertical relationships shows that there was a moderate difference in the mean rating by sector (forestry-sugarcane), $Pr(T < t) = 0.0266$. This means that transactional costs such as delayed payments from millers to producers and information search for production costs were perceived to have been more of a challenge in the sugarcane sector than the forestry sector. This finding was supported by qualitative data suggesting weaker collaborative relationships between millers and growers in the sugarcane sector, as per quotations below:

"Enough money and/or cyclic revenue is needed to run the business but KSL has a tendency of late payments going between 60-90 days. The delayed payment causes unnecessary interest accruals resulting into marginal profits" (Respondent, Sugarcane Sector).

"Initially we used to give cash advances to purchase plantations, and provided transport. However, the system was abused whereby some suppliers diverted the funds into other businesses... currently, we pay them within five working days after delivery to enhance their cash flow and introduced suppliers to Eco-bank for loan access. Right now the suppliers are self-sufficient, they can support themselves" (Plant Manager, Forestry Mill).

"The main challenge is the continuous reshuffle of ministers; before a minister gets acquainted with the industry another one is appointed. Even now the permanent secretaries are being transferred. At one time we broke down the costs to Minister Mukwaya. The minister requested the miller to give her the breakdown, but the miller refused. Recently, another meeting was organised with the Ministry of Trade, Industry and Cooperatives (MTIC) involving both out-growers and millers. The out-growers gave their cost breakdown of approx. 60 percent but the miller declined to give a cost breakdown" (Respondent, Association Executive Member and Opinion Leader (Sugar)).

The quotations above revealed that cost of searching for production costs of the value chain did not only impact upon transactional costs but also the reluctance by millers to reveal their production costs suggested possibility of opportunistic behaviors.

Our observations therefore are that some level of vulnerability to transactional costs between millers and growers in the sugarcane sector exits, compared to the forestry sector. However, similar results were obtained in the horizontal relationships in the forestry sector in comparison to the sugarcane sector. Using two-sample t-test, a comparison of the mean value for transactional costs in horizontal relationships shows that there was a slight difference in the mean rating by sector (forestry-sugarcane), $Pr(T > t) = 0.0548$. This means that there was suggestive evidence of minimum occurrence of transactional costs among producers in the forestry sector than the sugarcane sector. This quantitative finding was validated by the qualitative data finding below, which suggested occurrence of transactional costs as a result of replacing labor force taken by another grower without the consent of the labor force owner:

"I think we trust each other because we have same common ground. I have not seen many conflicts in client meetings, except one time a farmer accused his neighbors of stealing his workers." (Respondent, Forestry).

Therefore, on the question of the impact on performance differences between commercial sugarcane and forestry sector value chains regarding coordination of transaction costs within vertical and horizontal linkages, we concluded as follows. Moderate transactional costs such as information search costs and delayed payments from millers to producers were more of a challenge in the sugarcane sector than the forestry sector. This finding supported existence of weaker collaborative relationships between millers and growers for building production capabilities in the sugarcane sector. In the forestry sector, quantitative data revealed evidence of minimum occurrence of transactional costs among producers compared to the sugarcane sector. This quantitative finding was corroborated by the qualitative data finding which suggested occurrence of transactional costs as a result of replacing labor force taken by another grower without the consent of the labor force owner.

The above findings demonstrate that both vertical and horizontal collaborative relationships between primary producers and millers and among primary producers in the sugarcane and forestry industry value chains were vulnerable to some levels of transactional costs. Evidence of transactional costs suggests existence of opportunistic behaviors [4, 27]. Opportunistic behaviors were investigated by this study in the next question below.

Research question 3: How do actors' behaviors in vertical and horizontal linkages explain performance differences between commercial sugarcane and forestry sector value chains?

Vulnerability to transactional costs is mainly attributed to opportunistic behaviors. A comparison of the mean opportunism in vertical relationships revealed that there was a slight difference in the mean rating by sector (forestry-sugarcane), $Pr (T < t) = 0.0712$. This means that there was suggestive evidence of opportunism causing mistrust between millers and producers in the sugarcane sector than the forestry sector. This finding was validated by qualitative data findings below:

"Fairly good trust... however there is lack of transparency on the weigh bridge" (Association Chairman, Sugarcane Sector).

This above quotation was supported by the miller's representative quotation below suggesting some level of opportunistic behaviors along the sugarcane value chain.

"Small farmers are mainly the problem because at times they sell the fertilisers, but with commercial farmers, this is not quite rampart" (Agricultural Engineering Manager, Sugar Mill).

Generally, the finding of suspicious opportunistic behaviors between growers and millers validates the existence of weak collaborative relationships for building supplier production capabilities. On the other hand, a comparison of the mean value for opportunism in the horizontal relationships shows that there was no significant difference in the mean rating by sector (forestry-sugarcane), $Pr (T > t) = 0.1380$. This meant that manifestations of opportunistic behaviors were not quite rampart among producers in both the forestry and sugarcane sectors.

Therefore, in answering the question on how actors' behaviors in vertical and horizontal linkages explain performance differences between commercial sugarcane and forestry sector value chains, we observed the following. Quantitative data revealed suggestive evidence of opportunistic behaviors causing mistrust between millers and producers in the sugarcane sector than the forestry sector. This finding was validated by qualitative data that suggested the source of mistrust being due to lack of transparency of the miller's weighbridge and possibility of diverting inputs by growers to other farm activities rather than sugarcane growing.

The above findings show that vertical collaborative relationships between primary producers and millers in the sugarcane industry value chains were characterized by suspicion, thus affecting mutual trust as compared to the forestry sector value chain. This finding is in agreement with similar studies, which found out that mutually benefiting relationships develop trust [14, 28], while exploitative relationships exhibit low levels of trust and tend to be characterized by tensions that affect productivity gains in the value chains [29, 30, 31]. This was confirmed to be accurate inside Uganda's sugarcane sector value chain. Under these conditions, the GVC theoretical framework acknowledges that restructuring the governance structures of the value chain reduces opportunistic behaviors that contribute to susceptibility to transactional costs. Our research suggests that Uganda's sugarcane primary producers could enhance their market influence by transitioning from a captive value chain governance structure with high levels of dependency to either a modular or relational value chain governance structure with lower levels of dependency.The study found that opportunistic behaviors were not widespread in the horizontal interactions among primary producers in the forestry and sugarcane sectors. This discovery suggests that there are chances for primary producers, particularly in the sugarcane industry, to collaborate on investment strategies to reduce opportunistic behaviors that lead to transactional costs in vertical connections.

4. CONCISE SUMMARY

The study's main results showed that the sugarcane sector had a higher level of investment asset specificity in vertical connections between primary farmers and millers compared to the forestry sector. This discovery indicated the presence of effective collaborative relationships between suppliers and producers to improve productivity gains. A detailed analysis of the qualitative data revealed that production capabilities exist in the vertical connections, primarily through the utilization of contractors instead of the cane growers themselves. In the forestry sector, there was a more robust collaborative interaction between producers and producer support organizations for accessing inputs, information, and skills transfer compared to the sugarcane sector. The finding was supported by qualitative data and confirmed by observatory field data, validating the development of production capacity among primary producers in the forestry sector to increase productivity. The study found that controlling the spread of knowledge and skills transfer to primary producers indirectly through contractors allowed the miller(s) to consistently generate higher profits by providing low commodity prices, inputs, and services at high prices to sugarcane primary producers while maintaining weak supplier production capabilities. Opportunistic behaviors led to more suggestive evidence of transactional costs between millers and growers in the sugarcane sector than in the forestry sector. The important findings provided possible explanations for the performance discrepancies in reaching industry productivity benchmarks between sugarcane and forestry sector primary producers. Study findings revealed that a mere 18% of farmers reached the target industry productivity of 100 t/ha in their cane fields, suggesting that the majority (82%) of cane growers fell short of the projected industry productivity level. The forestry sector survey showed that 41.3% of farmers met industry performance standards, while 58.7% fell short of expectations.

5. AT A GLANCE

This study used an explanatory case study to assess supplier production capacities in order to improve productivity improvements in the value chains of Uganda's commercial forestry and sugarcane sectors. Key survey findings revealed that 18% of sugarcane farmers reached the target industrial productivity level of 100 t/ha, while the rest (82%) fell short of this goal. 41.3% of farmers in the forestry sector met industry performance standards, while 58.7% fell short of expectations. The behavior of large purchasers in developing suppliers by sharing knowledge, skills, and technology in collaborative value chain connections explains this paradox. Sugar millers in the sugarcane industry employed contractors to spread knowledge and skills, which diminished the production capacities of suppliers. Productivity in the forestry industry increased with the assistance of development partner agencies by effectively spreading information, skills, and suitable technology to primary producers. This discovery emphasizes the necessity of enacting intentional supplier development strategies by development partner agencies and governments to enhance productivity improvements in agri-business value chains in underdeveloped nations.

REFERENCES

1. WTO. Global Value Chains in a Changing World. Geneva, Switzerland: WTO Secretariat; 2013
2. Gereffi G, Humphrey J, Sturgeon T. The governance of global value chains. Review of International Political Economy. 2005;12(1):78-104
3. Smith A. An Inquiry into the Nature and Causes of the Wealth of Nations. 1776. Available: www./biblio.org/ml/libri/s/SmithA_WealthNations_pdf [Accessed: 24-02-2016]

4. Williamson EO. Markets and Hierarchies: Analysis and Antitrust Implications. New York: Free Press; 1975

5. Humphrey J and Schmitz H. Governance and Upgrading: Linking Industrial Cluster. And Global Value Chain Research. IDS Working Paper 120; 2000. https://www.ids.ac.uk/files/Wp120.pdf

6. Porter EM. Clusters and the new economics of competition. Harvard Business Review. 1998;76(6):77-90

7. Porter EM. Clusters and Economic Policy: Aligning Public Policy with the New Economics of Competition. Institute for Strategy and Competitiveness, Harvard Business School; 2009. Available: http://www.hbs.edu/faculty/PublicationFiles/Clusters_and_Economic_Policy_White_Paper_8e844243-aa23-449d-a7c1-5ef76. [Accessed: 30-07-2016]

8. Navdi K and Halder G. Local clusters in global value chains: Exploring dynamic linkages between Germany and Pakistan. Entrepreneurship & Regional Development. 2005;17:339-363

9. Martin R and Sunley P. Deconstructing clusters: Chaotic concept or policy panacea? Journal of Economic Geography. 2003;3:5-35

10. Eisenhardt KM. Building theories from case study research. Academy of Management Review. 1989;14(4):532-550

11. Birnberg JG, Shields MD, Young SM. The case for multiple methods in empirical management accounting research. Journal of Management Accounting Research. 1990;2:33-66

12. Atkinson AA, Balakrishnan R, Booth P, et al. New directions in management accounting research. Journal of Management Accounting Research. 1997;9:79-108

13. Shields MD. Research in management accounting by North Americans in the 1990s. Journal of Management Accounting Research. 1997;9(1):3-61

14. Chivaka R. Value creation through strategic cost management along the supply chain [PhD thesis]. South Africa: University of Cape Town; 2003

15. USCTA. Fourteenth Annual Report for Calendar year 2011. 2012. Available: www.ugandasugar.org. [Accessed 15-04-2015]

16. SPGS. 10 Years of Supporting Commercial Tree Planting in Uganda. No. 39, October–December. 2014. Available: www.sawlog.ug. [Accessed 15-04-2015]

17. USMA. The Seventeenth Annual Report for Calendar Year 2014. 2015. Available: www.ugandasugar.org [Accessed 15-04-2015]

18. Hurly KM, Sibiya TG, Nicholson R, King M. Road map for small-scale grower sustainability. Proceedings of the Annual Congress-South African Sugar Technologists' Association. 2015;88:318-336

19. Humphrey J, Schmitz H. Governance in global value chains. IDS Bulletin. 2001;32(3):19-29

20. Schimitz H. Value Chain Analysis for Policy Makers and Practitioners. UK: International Labour Organization, Institute of Development Studies, University of Sussex; 2005. Available: www.fao.org/fileadmin/user-upload/fisheries/docs/Value-Chain-ILO.pdf. [Accessed 15-04-2015]

21. UNIDO. Technological Upgrading in Global Value Chains and Clusters and their Contribution to Sustaining Economic Growth in Low and Middle Income Economies. Working Paper 3, Vienna: Italy. 2015. Available: http://www.unido.org//fileadmin/user_media/Services/PSD/WP_2015_03_v2.pdf. [Accessed: 02-08-2016]

22. WB. Growing Africa: Unlocking the Potential of Agri-Business. Washington, DC: The World Bank; 2013

23. Maya-Ambia JC. Constructing agro-industrial clusters or disembedding of the territory? Lessons from Sinaloa as the leading horticultural export-oriented Region of Mexico. The Open Geography Journal. 2011;4:29-44

24. Aznar-Sánchez AJ, Galdeano-Gómez E. Territory, cluster and competitiveness of the intensive horticulture in Almería (Spain). The Open Geography Journal. 2011;4:103-114

25. Giulian E, Pietrobelli C, Rabelloti R. Upgrading in global value chains: Lessons from Latin American clusters. World Development. 2005;33(4):549-573

26. Meyer-Stemer J, Maggi C, Seibel S. Upgrading in the tile industry of Italy, Spain and Brazil: Insights from cluster and value chain analysis. In Hubert S, editor. Local Enterprises in the Global Economy: Issues of Governance and Upgrading. Cheltenham: Edward Elgar Publishing; 2004

27. Williamson EO. Political institutions: The neglected side of the story–Comment. Journal of Law, Economics and Organization. 1990;6:263-266

28. Dyer HJ. Effective interim collaboration: How firms minimize transaction costs and maximise transaction value. Strategic Management Journal. 1997;18(7):535-556

29. da Silva BAC. The Growing Role of Contract Farming in Agri-Food Systems Development: Drivers, Theory and Practice. Rome: FAO; 2005. Available: http://www.fao.org/fileadmin/user_upload/ags/publications/AGSF_WD_9.pdf. [Accessed: 30-01-2016]

30. Brown O, Sander C. Supermarket Buying Power: Global Supply Chains and Smallholder Farmers. International Institute for Sustainable Development. 2007. Available: http://www.iisd.org/sites/default/files/publications/tkn_supermarket.pdf. [Accessed: 07-01-2013]

31. Tijaja JP. Exogenous factors and domestic agency in value chain dynamics: Lessons from the thai cassava value chains [PhD Thesis]. UK: The Open University; 2010

32. Frederick S, Gereffi G. 'Value Chain Governance'. United States Agency for International Development, Briefing Paper. 2009. Available: http://www.cggc.duke.edu/pdf/frederick_gereffi_valuechaingovernance_usaid:briefingpaper_feb2009

33. Schmitz H. Learning and earning in global garment and footwear chains. The European Journal of Development Research. 2006;18(4):546-571

13 AN ANALYSIS OF SUPPLY CHAIN PRICES FOCUSING ON PERISHABLE MARKETS

1. INTRODUCTION: AN OVERVIEW

Interest in price dynamics in perishable markets has significantly grown in recent decades. There has been a notable increase in the number of studies on this topic. The implications for agricultural markets, entrepreneurial strategies, and producer and consumer welfare are significant and warrant furtherinvestigation[1, 2, 3, 4, 5, 6, 7]:

The primary paradigm for analyzing price dynamics is the Law of One Price (LOP), which asserts that prices in distinct markets typically vary by an amount no greater than the shipping expenses involved in transferring a product from one market to another. Stiglitz [8] provides a formal definition stating that in the market, there is a consistent price and any price discrepancies are rapidly resolved through arbitrage opportunities. When price spreads are higher than transaction costs, arbitrageurs will intervene to narrow the spread, causing prices to adjust towards equilibrium.[1] Prices along the supply chain are interconnected through enduring connections and typically vary based on marketing expenses, which refer to the expenditures associated with promoting and selling the product. Simply said, retail prices are higher than farm pricing due to the extra expenses involved in marketing the product. Although it may seem straightforward in explanation, it is actually quite intricate in practice. Price dynamics are chaotic due to market movements, entrepreneurial strategies, physical restrictions, and biological dynamics. Examining pricing changes throughout the supply chain at various stages presents an interesting problem. The interest in understanding these characteristics has significantly grown in the recent few decades. Distinguishing between the evolution of marketing margins and the vertical transmission of prices is crucial. Marketing margins represent the price difference between different stages of the supply chain, while vertical price transmission focuses on how prices change and shocks are passed down the supply chain.This chapter examines research on price analyses, specifically focusing on studies related to perishable food markets such as fruits and vegetables. We examine research on marketing margins and vertical price transmission, and then go into key concerns concerning marketing margins and vertical price transmission in perishable goods markets. The remainder of the chapter is as follows: Section two discusses marketing margins, the following section covers vertical price transmission, section four analyzes marketing margins and vertical price transmission in perishable marketplaces, and the paper concludes with remarks and reflections for future studies.

2. ON MARKETING MARGINS

The terms marketing margin refer to the difference between the retail price, paid by consumers for a finished product, and the farm price of raw products. For instance, assuming raw tomatoes are sold at 1.5 €, the marketing margin for processed tomatoes, say sold at 2 €, would be the difference of the two prices (in our example the marketing margins equals 0.5 €). Clearly the importance of analyzing marketing margins relies on the possibilities to analyze how market equilibria evolve at different stages of the supply chain.

Myers et al. [10] review a century of research on marketing margins, starting from the seminal analyses of Waugh [11] based on descriptive statistics aimed at assessing the size of marketing margins, and concluding with the description of Schroeter and Azzam [12] and Holt [13] on the impact of risk on marketing margins.

Wohlgenant [14] provides a complete survey on theoretical and empirical issues in marketing margins analysis underlying some of the questions of interest for researchers and policymakers (e.g., are margins too large/small compared to farm and retail prices? Why they differ among products/space/time? How quickly are price shocks transmitted along the supply chain? What are the determinants of margins movements? etc.).

The basic model to analyze marketing margins is a two-market static model subject to demand and supply shifters (i.e., exogenous variables capable of increasing or decreasing the whole market demand, or market supply). It is generally used for comparative statistics [15, 16] as it allows to conclude on margins dynamics under different economic assumptions (e.g., fixed and variable input proportions, markup pricing, etc.). However, Gardner [15, p. 406] pointed that "no simple markup pricing rule can depict the relation between the farm and retail prices," and generally, the empirical approaches lack of theoretical foundations.

Wohlgenant [14] revises the possible factors affecting marketing margins: an extensive literature has focused on the role of market power in order to provide theoretical [12, 17, 18, 19, 20] and empirical evidences [21, 22]; other factors that affect farm and retail prices spread are price risks [12, 13], technical and structural changes [23, 24], product quality and seasonality [25, 26]. Wohlgenant' survey concludes that the approaches to model marketing margins are still inappropriate since they ignore significant economic aspects (namely the input substitutability between farm input and other inputs in producing retail products), and more research is needed either to understand the role of the mentioned factors on marketing margins as well as the role of actual trends in agricultural sector (e.g., increasing vertical integration and coordination, growing expenses in advertisements, introduction of new food safety regulations, etc.).

Traub and Jayne [27] investigated how price (de)regulation influenced the size of the marketing margins of maize ion South Africa. They found that deregulation leads to an increase in marketing margins of 20%: the authors admit their results are not supported by the literature and may prove to be disruptive of existing evidence. More recently, Dawe and Maltsoglou [28] investigated how price increases affect the welfare impacts, and what is the influence of marketing costs (i.e., the costs to market the goods). They show the importance of assuming a correct functional form for marketing costs and found that in a vast majority of cases, it is safer to assume that marketing costs are fixed, rather than proportional.

3. ON VERTICAL PRICE TRANSMISSION

Meyer and Cramon-Taubadel [29] present the state of the art of the literature on asymmetric price transmission discussing on the theories and empirical aspects. The adjustment to price shocks along the supply chain, from producer to wholesale and retail levels, and vice-versa, is an important aspect of the functioning of supply chains. Asymmetries in vertical price transmission may be due to several aspect of the market structure. For instance, imperfect price transmission may be caused by market power that induces oligopolistic behavior. The consequences of vertical price transmission are worth investigation: in fact, the asymmetry and the speed of price transmission from farm level to final consumers result in positive or negative welfare effects for economic agents.

The literature on vertical price transmission is focused on four fundamental topics [30]: the magnitude of how price shocks are vertically transmitted along the supply chain; the speed of transmission of such shocks; the symmetric or asymmetric nature of price transmission; the direction of price transmission in terms of whether shocks are transmitted upwards or downwards.

According to Meyer and Cramon-Taubadel [29], there are two main causes of asymmetries in vertical price transmission: imperfect competition (i.e., market power) and adjustment costs. Moreover, asymmetries in price transmission seem to be related to political interventions, asymmetric information and inventory management.

Bailey and Brorsen [31] argue that no a priori explanations arising from the degree of market power may help predicting the positive or negative nature of asymmetries in price transmission: several authors [32, 33, 34, 35, 36] conclude that market power may induce asymmetric transmission, with positive asymmetric price transmission being induced by monopolistic behavior (i.e., increases in input prices tend to squeeze marketing margins; moreover, decreases in output prices are likely to be transmitted faster and/or more completely than increases in output prices). Lately, McCorriston et al. and Lloyd et al. [37, 38, 39] develop a framework to show how market power may lead to imperfect price transmission. Indeed, also in oligopoly, both positive and negative asymmetric price transmission may occur. Summing up, the literature still lacks of a solid link between market power and asymmetry in price transmission [29].

Another major explanation for asymmetric price transmission (APT) is provided by asymmetric adjustment costs[2] arising when firms change the quantities and/or prices of inputs and/or outputs. Bailey and Brorsen [31] conclude that positive asymmetric price transmission may be induced by the easiness for firms facing output reduction to disemploy inputs rather than to augment production by recruiting new inputs. Differently, Ward [40] concludes that negative asymmetric price transmission is likely to occur in markets of perishable products: retailers tend to hesitate to raise prices as they fear potential reductions in sales due to wastes for spoilage. Heien [41] raises arguments against Ward's conclusions: he argues that changing prices is less problematic when dealing with perishable products rather when dealing with storable products in that the price adaptation for products with long shelf-life requires high time costs and losses of goodwill. Finally, Peltzman [42] finds no evidence of a relationship between menu costs and APT, which might rather depend by the presence of high menu costs supported in fragmented supply chains. The strategic management of inventories may help to adapt production to exogenous shocks: as a result, the managerial strategies on inventories have been mentioned as possible cause of asymmetric price transmission. Balke et al. [43] argue that accounting methods such as first-in-first-out may lead to asymmetric price transmission. Blinder [44] and Reagan and Weitzman [45] argue that inventory management leads to positive APT. In particular, Reagan and Weitzman suggest that in periods of low demand firms will adjust the quantity produced and increase inventory rather than decrease output prices, increasing prices during periods of high demand. In summary, also for adjustments costs, the conclusions are ambiguous and sometimes contradictory, with some authors providing arguments for positive, and others for negative APT.

Another aspect that merit attention is the potential impact of government interventions, in terms of producer subsidies. Gardner [15] concludes on their role in influencing the asymmetries in farm-to-retail price dynamics, and evidence on such a mechanism is provided in the diary sector [6, 46]. Bailey and Brorsen [31] conclude on the role of asymmetric information in determining APT and point out that asymmetries in price series data can result from a distorted price reporting process. Kinnucan and Forker [6] and Cramon-Taubadel [47] consider APT in the framework of the Gardner's marketing margin model: the price spread between farm and retail levels is due to demand changes at retail-level as well as to changes in supply at farm level. Assuming perfect competition and constant returns to scale, the model suggests that changes in demand at retail level may have a large influence on marketing margins with respect to changes in supply at farm level. Kinnucan and Forker [6] argue that the marketing margins induced by these shifts tend to provoke asymmetric price transmission; conversely, Cramon-Taubadel [47] concludes that positive or negative APT arises depending on the predominant shift.

In conclusion, although many studies found imperfect price transmission along the supply chain, there is not a clear consensus but rather a variety of evidences depending on commodities, countries and data under analysis. It seems evident that more research relationships among prices along the supply chain and the underlying behavior of agents are needed: Although a large number of studies have investigated the phenomenon of price transmission in agricultural markets, it is still not possible to draw strong conclusions to support policy decisions. Meyer and Cramon-Taubadel [29] are skeptical on the actual results provided by the literature unable to explain the economical relevance of evidence of imperfect price transmission. They suggest that it would be premature to draw far-reaching conclusions for theory and policy on the basis of work to date: their critic pertains either the commonly adopted tests to measure the transmission, still not fully reliable and precise in the statistical sense, and the theories capable to explain asymmetric price transmission.

More recently, there is been a renewed turmoil in the literature that leads to an increasing number of studies on vertical price transmission. First, Frey and Manera [48] reviewed the literature on econometric models of asymmetric price transmission, concluding that asymmetries are likely to occur in a wide range of markets. The authors deepen on the sets of models adopted to study asymmetries: autoregressive distributed lags, partial adjustments, error correction models, and vector autoregressive models. Clearly, more recent models are not surveyed in the mentioned study. For instance, Brummer et al. [49] adopt a Markov switching vector error correction model to investigate asymmetries in the wheat market. Hassouneh et al. [50] adopt a STAR model (smoothing transition autoregressive model) to conclude on vertical price transmission in the poultry sector.

Several other papers have instigated price dynamics along the supply chain: Acosta and Valdes [51] and Antonioli and Santeramo [52] explore price dynamics in diary markets; Santeramo and Cramon-Taubadel [53] focus on perishable food products; Tifaoui and Cramon-Taubadel [54] investigate the dynamics in butter market. The list may be long and never exhaustive as more and more papers are published every year. A recent survey that is worth reading has been recently published by Swinnen and Vandeplas [55]: they deepen on conceptual issues in price transmission analyses of agricultural supply chains.

4. A FOCUS ON PERISHABLE MARKETS

Marketing margins and vertical price transmission in fresh produce markets have been objects of studies due to the relevant policy implications deriving from such analyses (e.g., the assessment of the efficiency of the produce marketing system and/or of the functioning of markets both vertically and spatially separated), but nowadays, the literature still lacks of economic explanations for the peculiar results found for perishable products [56]. The present section briefly reviews the findings of recent studies.

Wohlgenant [14] estimated an econometric model to assess marketing margins dynamics in eight commodities sectors detecting symmetric dynamics, compatible with a competitive behavior, in all but fresh fruits sector, for which he found evidence of constant return to scale but not for competitive behavior. These findings have important implications for supply chain. In particular, the evidence of constant return to scale for perishable products (e.g., fruits) suggests that competitive behavior is impeded by perishability. In other terms, the fear of losing products due to waste for spoilage prevents the adoption of competitive pricing strategies along the supply chain.

As regard price transmission,[3] Ward [40] seminal work on vertical price transmission in perishable goods markets determined the supply chain ring primarily responsible for establishing price and also tested for pricing asymmetries. He found that the pricing point for fresh produce existed at the wholesale market level (price transmission runs from wholesale markets to retail and producer levels) and that price decreases were more likely to be fully passed on to the retail and producer level sectors than were price increases. As above mentioned, he argued that retailers selling perishable goods might be reluctant to raise prices in line with an increase in farm-level prices given the risk that they will be left with unsold spoiled product. Heien [41] raises different arguments and concludes that changing prices is less problematic when dealing with perishable products. Although the mentioned papers are dated, not many theoretical advances have been made, nor the controversial theories abovementioned have been clarified.

Girapunthong et al. [57] applied Ward's pricing asymmetry model to fresh tomatoes data: he found the prices at producer level influence price at wholesale and retail levels; he also argues that increases in producer prices have a major impact on wholesale prices than decreases in producer prices. The controversy results might be explained by the structural changes occurred in the market for fresh tomatoes, that is, by changes in the entire structure of the market (e.g., changes in amount of contracts, concentration of retailers and suppliers, etc.).

Table 1. –Major findings in applied analyses of vertical price transmission in perishable markets.

Author	Journal	Year	Product	Frequency	Results[a]
Aguiar and Santana	Agribusiness	2002	Tomatoes	Monthly	Positive asymmetry
			Onions	"	Symmetry
Bakucs et al.	Studies in Agricultural Economics	2007	Potatoes	Monthly	Symmetry
			Carrots	"	Symmetry
			Parsley	"	Symmetry
			Tomatoes	"	Positive asymmetry
			Peppers	"	Symmetry
Bernard and Willett	Journal of Agricultural and Applied Economics	1996	Broiler	Monthly	Negative asymmetry
Bernard and Willett	Applied Economics Letters	1998	Broiler	Weekly	Symmetry
			Broiler	Monthly	Positive asymmetry
Brooker et al.	Journal of Food Distribution Research	1997	Peppers	Weekly	Positive asymmetry
Hassan and Simioni	ÉconomieRurale	2004	Tomatoes	Weekly	Negative asymmetry
			Chicory	"	Negative asymmetry
Girapunthong et al.	Journal of Food Distribution Research	2003	Tomatoes	Monthly	Asymmetryb
Hassan and Simioni	ÉconomieRurale	2004	Chicory	Weekly	Symmetry
			Tomatoes	"	Symmetry
Heien	American Journal of Agricultural Economics	1980	Potatoes	Monthly	Positive asymmetry
			Apples	"	Symmetry
			Oranges	"	Negative asymmetry
			Lettuce	"	Symmetry
			Tomatoes	"	Symmetry
Kuiper and Lansink	Agri Business	2013	Broiler	Monthly	Positive asymmetry
			Apples	Monthly	Symmetry
			Carrots	"	Symmetry
			Potatoes	"	Symmetry

Author	Journal	Year	Product	Frequency	Results[a]
Pick et al.	Agri Business	1990	Lemons	Weekly	Positive asymmetryc
			Oranges	"	Positive asymmetryc
Powers	Agri Business	1995	Lettuce	Weekly	Positive asymmetry
Schertz Willet et al.	Agri Business	1997	Apples	Monthly	Positive asymmetry
Ward	American Journal of Agricultural Economics	1982	Carrots	Monthly	Symmetry
			Celery	"	Negative asymmetry
			Cabbage	"	Negative asymmetry
			Cucumbers	"	Symmetry
			Peppers	"	Negative asymmetry
			Potatoes	"	Negative asymmetry
			Tomatoes	"	Negative asymmetry
Worth	Economic Research Service	1999	Carrots	Monthly	Positive asymmetry
			Celery	"	Symmetry
			Lettuce	"	Symmetry
			Onions	"	Symmetry
			Potatoes	"	Symmetry
			Tomatoes	"	Positive asymmetry

[a]Results on symmetry, positive and negative asymmetry depend on time frequency.

[b]Positive asymmetry among wholesaler and retailer prices; negative asymmetry among wholesaler and producer prices.

[c]However, over time price changes appear to be symmetric.

Sexton et al. [58] completed a study to analyze grocery retailer behavior concluded that there is considerable independence in retailers in setting prices for produce commodities and that higher volumes led to larger margins. Similarly, Girapunthong et al. [57] found that the retail price is more likely to change after increases in producer prices than after decreases in producer prices. According to Sexton et al. [58], retailers have been more aggressive in using their market power to influence prices paid to producers for product (e.g., higher volumes of product in the market are used to bid down producer prices without equal declines in retail prices, widening the farm-to-retail margin), and many shippers tried to counter the market power exercised by large retailers consolidating their businesses with firms located in several areas of production.

Recently, Santeramo and Cramon-Taubadel [53] show that vertical price transmission is symmetric for products not affected by large losses for spoilage and tends to be asymmetric for very perishable products. Their results are in line with numerous studies [40, 42, 46] and in contrast with the results presented by Kim and Ward [59]. In addition, Santeramo [7] concludes that, in markets of perishable products, price rises tend to be slowly transmitted and do not influence distant markets; on the contrary, shocks originated by price decreases spread across markets. The results are in line with Ward [40] who has demonstrated that the high perishability, and the inability to delay sales through temporary storage implies that decreases in wholesale prices have a larger effect on retail prices than increases.

5. CONCISE SUMMARY

Interest in price dynamics in perishable markets has significantly grown in recent decades. There has been a notable increase in the number of studies on this topic. The implications for agricultural markets, entrepreneurial strategies, and producer and consumer welfare are significant and warrant further investigation[1, 2, 3, 4, 5, 6, 7].The primary paradigm for analyzing price dynamics is the Law of One Price (LOP), which asserts that prices in distinct markets typically vary by an amount no greater than the shipping expenses involved in transferring a product from one market to another. Stiglitz [8] provides a formal definition stating that in the market, there is a consistent price and any price discrepancies are rapidly resolved through arbitrage opportunities. When price spreads are higher than transaction costs, arbitrageurs will intervene to narrow the spread, causing prices to adjust towards equilibrium.1 Prices along the supply chain are interconnected through enduring connections and typically vary based on marketing expenses, which refer to the expenditures associated with promoting and selling the product. Simply said, retail prices are higher than farm pricing due to the extra expenses involved in marketing the product. Although it may seem straightforward in explanation, it is actually quite intricate in practice. Price dynamics are chaotic due to market movements, entrepreneurial strategies, physical restrictions, and biological dynamics. Examining pricing changes throughout the supply chain at various stages presents an interesting problem. The interest in understanding these characteristics has significantly grown in the recent few decades. Distinguishing between the evolution of marketing margins and the vertical transmission of prices is crucial. Marketing margins represent the price difference between different stages of the supply chain, while vertical price transmission focuses on how prices change and shocks are passed down the supply chain.This chapter examines research on price analyses, specifically focusing on studies related to perishable food markets such as fruits and vegetables. We examine research on marketing margins and vertical price transmission, and then go into key concerns concerning marketing margins and vertical price transmission in perishable goods markets.

The remainder of the chapter is as follows: Section two discusses marketing margins, the following section covers vertical price transmission, section four analyzes marketing margins and vertical price transmission in perishable marketplaces, and the paper concludes with remarks and reflections for future studies.

6. AT A GLANCE

Prices at various stages of the supply chain are interconnected by enduring ties and typically vary due to marketing expenses. Various factors contribute to the complexity and unpredictability of pricing fluctuations throughout the supply chain. Debates persist regarding the evolution of marketing margins over time and across different commodities, as well as the transmission of pricing along the supply chain. Scholars have been compelled to pay special attention to the implications for the economy, company management, and market laws. The evolution of prices in the supply chain of perishable food is a challenging topic that has sparked a heated debate. We examine existing research on pricing analysis, marketing margins, and vertical price transmission, focusing on perishable products such as fruits and vegetables, to identify unresolved issues that warrant additional study.

REFERENCES

1. Santeramo FG, Cioffi A. Spatial price dynamics in the EU F&V sector: the cases of tomato and cauliflower. Presented at the 116th Seminar, October 27-30, 2010, Parma, Italy 95228. European Association of Agricultural Economists
2. Santeramo FG, Cioffi A. The entry price threshold in EU agriculture: Deterrent or barrier? Journal of Policy Modeling. 2012;34(5):691-704
3. Santeramo FG, Cioffi A. Transmission of market crises in the European vegetables sector. Politica Agricola Internazionale-International Agricultural Policy. 2012;2012(2):37-46
4. Santeramo FG. Market integration in European Union's fruits and vegetables sector. Unpublished PhD dissertation. University of Naples "Federico II". 2010
5. García-Germán S, Bardají I, Garrido A. Evaluating price transmission between global agricultural markets and consumer food price indices in the European Union. Agricultural Economics. 2016;47(1):59-70
6. Kinnucan HW, Forker OD. Asymmetry in farm-retail price transmission for major dairy products. American Journal of Agricultural Economics. 1987;69:307-328
7. Santeramo FG. Price transmission in the European tomatoes and cauliflowers sectors agribusiness: An International Journal. 2015;31(3):399-413
8. Stiglitz JE. Information. In: Henderson DR, editor. The Fortune Encyclopedia of

Economics. New York: Warner Books; 1993. pp. 16-21

9. Santeramo FG, Di Gioia L. On the analysis of price dynamics in agricultural markets. In: Rosa S, editor. Progress in Economics Research. Vol. 39. Nova Science Publishers. 2017
10. Myers RJ, Sexton RJ, Tomek WG. A century of research on agricultural markets. American Journal of Agricultural Economics. 2010;92(2):376-403
11. Waugh FV. Margins in marketing. Journal of Farm Economics. 1934;16(2):233-245
12. Schroeter JR, Azzam AM. Marketing margins, market power, and price Uncertainty. American Journal of Agricultural Economics. 1991;73:990-999
13. Holt MT. Risk response in the beef marketing channel: A multivariate generalized ARCH-M approach. American Journal of Agricultural Economics. 1993;75:559-571
14. Wohlgenant MK. Marketing margins: Empirical analysis. In: Gardner B, Rausser G, editors. Handbook of Agricultural Economics. Vol. 1. 2001. Chapter 16, pp. 934-970
15. Gardner BL. The farm-retail price spread in a competitive food industry. American Journal of Agricultural Economics. 1975;57:399-409
16. Fisher BS. The impact on changing marketing margins on farm prices. American Journal of Agricultural Economics. 1981;63:261-263
17. Schroeter JR. Estimating the degree of market power in the beef packing industry. The Review of Economics and Statistics. 1988;70:158-162
18. Azzam AM, Pagoulatos E. Testing oligopolistic and oligopsonisticbehaviour: An application to the U.S meat-packing industry. Journal of Agricultural Economics. 1990;41:362-370
19. Muth MK, Wohlgenant MK. A test for market power using marginal input and output prices with application to the U.S. beef processing industries. American Journal of Agricultural Economics. 1999;81:638-643
20. Lloyd T, McCorriston S, Morgan CW, Rayner AJ. The impact of food scares on price transmission in inter-related markets. Paper presented to the XXVth IAAE Conference in Durban/South Africa. 2003
21. Wohlgenant MK. Demand for farm output in a complete system of demand functions. American Journal of Agricultural Economics. 1989;71:241-252
22. Holloway GJ. The farm-retail price spread in an imperfectly competitive food industry. American Journal of Agricultural Economics. 1991;73:979-999
23. Hennessy DA. Microeconomics of agricultural grading: Impacts on the marketing

channel. American Journal of Agricultural Economics. 1995;77:980-989

24. Perrin RK. The impact of technological change on a competitive industry. Journal of Agricultural and Resource Economics. 1997;22(2):345-355. December
25. Parker DD, Zilberman D. University technology transfers: Impacts on local and U.S. Economies. Contemporary Economic Policy. 1993;11:87-99
26. Lyon CC. Thompson GD. Temporal and spatial aggregation: Alternative marketing margin models. American Journal of Agricultural Economics. 1993;75:523-536
27. Traub LN, Jayne TS. The effects of price deregulation on maize marketing margins in South Africa. Food Policy. 2008;33(3):224-236
28. Dawe D, Maltsoglou I. Marketing margins and the welfare analysis of food price shocks. Food Policy. 2014;46:50-55
29. Meyer J, Cramon-Taubadel S. Asymmetric price transmission: A Survey. Journal of Agricultural Economics. 2004;55(3):581-611
30. Vavra P, Goodwin BK. Analysis of Price Transmission Along the Food Chain. OECD Food. Agriculture and Fisheries Working Papers. No. 3, OECD Publishing; 2005
31. Bailey D, Brorsen BW. Price asymmetry in spatial fed cattle markets. Western Journal of Agricultural Economics. 1989;14:246-252
32. Boyd MS, Brorsen BW. Price asymmetry in the U.S. pork marketing channel. North Central Journal of Agricultural Economics. 1988;10:103-109
33. Karrenbrock JD. The behaviour of retail gasoline prices: Symmetric or Not? Federal Reserve Bank of St. Louis Review. 1991;73:19-29
34. Appel V. Asymmetrie in der preistransmission. AgrarwirtschaftSonderheft. 1992;135:178-213
35. Griffith GR, Piggot NE. Asymmetry in beef, lamb and pork farm-retail price transmission in Australia. Agricultural Economics. 1994;10:307-316
36. Mohanty S, Peterson EWF, Kruse NC. Price asymmetry in the International wheat market. Canadian Journal of Agricultural Economics. 1995;43:355-366
37. McCorriston S, Morgan CW, Rayner AJ. Processing technology, market power and price transmission. Journal of Agricultural Economics. 1998;49(2):185-201
38. McCorriston S, Morgan CW, Rayner AJ. Price transmission: The interaction between market power and returns to scale. European Review of Agricultural Economics. 2001;28(2):143-159
39. Lloyd TA, McCorriston S, Morgan CW, Rayner AJ. Food scares, market power

and price transmission: The UK BSE crisis. European Review of Agricultural Economics. 2006;33(2):119-147

40. Ward RW. Asymmetry in retail, wholesale and shipping point pricing for fresh vegetables. American Journal of Agricultural Economics. 1982;62:205-212
41. Heien DM. Markup pricing in a dynamic model of food industry. American Journal of Agricultural Economics. 1980;62:10-18
42. Peltzman S. Price rise faster than they fall. Journal of Political Economy. 2000;108:466-502
43. Balke NS, Brown SPA, Yücel MK. Economic Review. Dallas: Federal Reserve Bank of Dallas; (First Quarter) 1998. pp. 2-11
44. Blinder AS. Inventories and sticky prices: More on the microfoundation of macroeconomics. The American Economic Review. 1982;72:334-348
45. Reagan PB, Weitzman ML. Asymmetries in price and quantity adjustments by the competitive Firm. Journal of Economic Theory. 1982;27:410-420
46. Serra T, Goodwin BK. Price transmission and asymmetric adjustment in the Spanish dairy sector. Applied Economics. 2003;35:1889-1899
47. Cramon-Taubadel S. Estimating asymmetric price transmission with the error correction representation: An application to the German pork market. European Review of Agricultural Economics. 1998;25:1-18
48. Frey G, Manera M. Econometric models of asymmetric price transmission. Journal of Economic Surveys. 2007;21(2):349-415
49. Brümmer B, Cramon-Taubadel S, Zorya S. Vertical price transmission between wheat and flour in Ukraine: A markov-switching vector error correction Approach. European Review of Agricultural Economics. 2009;36:203-230
50. Hassouneh I, Radwan A, Serra T, Gil JM. Food scare crises and developing countries: The impact of avian influenza on vertical price transmission in the Egyptian poultry sector. Food Policy. 2012;37(3):264-274
51. Acosta A, Valdés A. Vertical price transmission of milk prices: are small dairy producers efficiently integrated into markets? Agribusiness. 2014;30(1):56-63
52. Antonioli F, Santeramo FG. Vertical price transmission in milk supply chain: Market changes and asymmetric dynamics. Paper Presented at the 6th AIEAA Conference. Piacenza. 2017
53. Santeramo FG. Cramon-taubadel S. On perishability and vertical price transmission: Empirical evidences from Italy. Bio-Based Applied Economics. 2016;5(2):199-214

54. Tifaoui S, Cramon-Taubadel S. Temporary sales prices and asymmetric price transmission. Agri Business. 2016;33(1):85-97

55. Swinnen J, Vandeplas A. Price transmission in modern agricultural value chains: Some conceptual issues. Food Price Dynamics and Price Adjustment in the EU. 2015;147:148-166

56. VanSickle J. Spatial and vertical price transmission in fresh produce markets. Presented at the Market Integration and Vertical And Spatial Price Transmission in Agricultural Markets Workshop at the University of Kentucky Lexington, KY. April 21, 2006

57. Girapunthong N, VanSickle JJ, Renwick A. Price asymmetry in the United States fresh tomato market. Journal of Food Distribution Research. 2003;34:51-59

58. Sexton RM, Zhang M, Chalfant J. Grocery retailer behavior in the procurement and sale of perishable fresh produce commodities. USDA Economic Research Service. Report No 2. September, 2003

59. Kim H, Ward RW. Price transmission across the US food distribution system. Food Policy. 2013;41:226-236

60. Kinnucan HW, Zhang D. Notes on farm-retail price transmission and marketing margin behavior. Agricultural Economics. 2015;46(6):729-737

61. Cioffi A, Santeramo FG, Vitale CD. The price stabilization effects of the EU entry price scheme for fruit and vegetables. Agricultural Economics. 2011;42:405-418

62. Cioffi A, Santeramo FG. The entry price threshold in EU F&V sector: deterrence or effective barrier? Presented at the 122nd Seminar, Ancona, Italy, European Association of Agricultural Economists. February 17-18, 2011

63. Stephens EC, Mabaya E, Cramon-Taubadel S, Barrett C. Spatial price adjustment with and without trade. Oxford Bulletin of Economics and Statistics. 2011;73:453-469

14 GENDERED ASPECTS OF PRIMARY VALUE CHAINS IN SOUTHWESTERN MOROCCO

1. INTRODUCTION: AN OVERVIEW

40% of Morocco's population relies on agriculture as a major income source, offering stability to disadvantaged farm households facing challenges like weather changes and desertification. In 2008, the government initiated Plan Maroc Vert (Morocco Green Plan (MGP)), a nationwide project aiming to transition agriculture policy from a focus on protection to market-oriented principles. The first pillar of the plan focuses on increasing value-added, productivity, and improving access to export markets for high-value commodities by making strategic investments to modernize the commercial agriculture sector. Pillar II emphasizes the significance of public support in assisting small rural agricultural households in marginalized areas, a critical sector since the 'Arab Spring' upheavals of 2011 [1]. This second pillar aims to improve the quality and stability of local commodity supply from resource-poor farm households by establishing production cooperatives within a federated cooperative structure, with a focus on socioeconomic factors.

Substantial sums have been allocated to enhance the quality of life for impoverished rural communities. These strategies consist of: (i) establishing producer cooperatives for women only, men only, or mixed groups; (ii) providing state subsidies for processing and packaging equipment, funding infrastructure development such as warehouses for product sorting and export preparation; (iii) organizing exhibitions and fairs to showcase local products globally; (iv) supplying primary inputs to cooperative members free of charge; and (v) implementing geographical indication markers to safeguard cultural heritage rights and gain economic advantages. Despite widespread attention in national and international media, the assertion that rural farm households are not fairly compensated for their highly valued products in both foreign and local markets remains prevalent. The authors conducted participatory workshops which uncovered a common belief in southern Morocco that marketing agents are responsible for unfair returns due to their exploitation of female illiteracy, ineffective production methods, and absence of formal credit systems in rural regions. Our discoveries present an alternative perspective that differs from commonly accepted beliefs. We contend that the absence of synchronization in public services, combined with existing cultural norms, policies, and environmental limitations, hinders the capacity for private and public endeavors focused on enhancing efficiencies in crucial (pro-poor) value chains to enhance sustainable (long-term) rural livelihoods.

This chapter focuses on analyzing the production and marketing systems of four native products in southern Morocco (argan, cactus, saffron, rose), with a particular emphasis on the gender equity difficulties present in each value chain. Each of the four possesses unique features that are highly sought after and valuable in local and international markets, particularly in specific regions and villages. These four products are grown on irrigated regions on the outskirts of conventional export commodities such as tomatoes, citrus fruits, and berries. [4]. They hold great economic value for rural populations and are increasingly gaining global acknowledgment. Argan, traditionally valued by the Amazigh tribes of Morocco, was recognized internationally in the late 1980s for its health and beauty benefits after scientific research revealed its positive attributes. Saffron production in the villages of Taliouine and Taznakht in southern Morocco ranks Morocco fourth in global saffron output due to the need for competent female labor to maintain quality.[5] The roses of KalaaM'Gouna hamlet generate highly prized rose oil, making Morocco one of the top producers of rose oil, historically ranking third after Bulgaria and Turkey. Prickly pear cactus in Sidi Ifni, a southern province, is highly sought after nationally in Morocco, commanding high prices in both wholesale and retail marketplaces. It also has the potential for a profitable worldwide market for fresh fruit and cosmetic oil.[6]. While much has been written on botanical and other scientific and technical aspects of these local products [7, 8, 9, 10, 11, 12], little attention has been given (at least in the English literature) to an analysis of policy, market and social challenges faced by small producers, and opportunities for addressing these challenges. What does exist [13, 14, 15] is restricted to argan and with significant contribution to a contemporary understanding of historical challenges that continue to persist today. Surprisingly, there is a dearth of research and attention to gender-specific concerns related to these commodities, despite the enormous economic contribution of women.

This study uses qualitative and participatory action research methods to explore gender-based participation in four local commodities. It focuses on trade-offs in time and income allocation, as well as household power dynamics. Trade-offs occur in various sectors, including health, education, farm and off-farm income prospects, and notably in terms of unequal gender-based access to resources and economic chances for work. An investigation of empowerment and meaning reveals a substantial danger to sustainability in economic empowerment and well-being. For a more in-depth analysis and full description of methodologies, along with a list of interviewed respondents, go to the working paper that this chapter is based on [16]. We argue that the Morocco Green Plan lacks consideration for gender concerns and must focus more on addressing gender equality in economic opportunities to meet its goals effectively.

2. MOROCCO GREEN PLAN

Plan Maroc Vert (Morocco Green Plan) is built on two pillars with four stated objectives:

1. Reduced poverty within rural areas
2. Improved measures for achieving national food security
3. Natural resource sustainability
4. Improved and equitable access to national and international markets.

Pillar I: One aim, within the early days of the initiation of the plan was to generate 7 billion Euros of investment, 75% of which was expected to be generated from private investment across 961 projects identified for implementation. A considerable number of these projects were targeted at value-added initiatives within irrigated production areas with high agricultural production potential and underpinned by a desire to 'modernize' the agricultural sector.

Pillar II: In addition to the activities under pillar I, an additional 545 projects were targeted for remote areas, particularly within marginalized rural areas where poverty continues to persist. These initiatives seek to uncover avenues for intensification, diversification, and specialization in agricultural production and processing activities through directed social and economic organization (production cooperatives and a federated model of cooperative organization).

Taken together, these two pillars are underpinned by a stated desire to address socioeconomic disparities between "modern" and "traditional" agricultural sub-sectors, through improvements in productivity (primary and processed), and with enhanced access to commercial markets.

3. CASE STUDIES: ARGAN, ROSE, CACTUS, AND SAFFRON VALUE CHAINS

3.1. Argan: A Story of Women Versus Machine

Endemic to Morocco's public forests in southwestern Morocco, argan maintains historical, cultural, and economic importance. For centuries, the argan tree has been associated with the Amazigh (Berber) tribes who have relied on its oil, and associated joint products, for culinary, cosmetic, and health purposes; feed and fodder for small ruminants; and as a source of heating material [17]. But the argan tree is now under threat, with implications for Amazigh tradition and customs. Oral discussions with officials and industry representatives indicate (without the support of publicly available data) that natural argan forests have shrunk from an estimated 1.4 million hectares at the turn of the century to an estimated 800,000 hectares today. While traditional systems of usufruct rights for harvesting argan fruit continue to exist, access to these forests is banned by the Ministry of Forestry for 3 months of the year, so as to mitigate the impact of grazing over the fruit-bearing season. Yet, despite oversight by the Ministry, grazing and illicit felling of argan trees for charcoal production are reported to be on the rise by local communities. Incidence of social tension between local residents and camel herders from the disputed Western Sahara, who graze in the argan forests, was mentioned during a number of discussions with community elders and with indication of a rising trend in violence.

One fundamental issue related to access rights and a cause for tension with nomadic camel and goat herders is the nature of ownership and usage rights for argan trees. Under existing legislation, historical rights to harvest argan trees have been maintained and passed down by inheritance. Within villages, communities and tribes, specific argan trees are held by households on the basis of cultural norms, and common knowledge exists on rights to the fruit from these trees. Yet, the trees themselves belong to the state as national heritage, even if they exist on private property; the incentive to plant argan trees on private property, therefore, is mitigated, as is any desire to maintain trees in public forests through pruning, management or general care in situ. Taken together, extended drought, excessive grazing, illegal felling, and issues of ownership have been blamed for the reduction in argan forest areas; yet, many also argue that an increasing price for argan oil has also led to an effective mopping up of argan fruit and seed from the forest floor, thereby leading to reduced natural regeneration.

Annual production of argan (fruit) and oil is neither accurately collected nor officially reported, but estimates in the late 1980s were on the order of 4000 tonnes of argan oil produced annually [18]. Argan oil is sold locally as a traditional source of edible oil and internationally as high-valued inputs into cosmetics, skin care products, and shampoos. The latter has taken on increasing importance since the 1980s, at which time scientific curiosity and evidence-based research revealed several positive cosmetic and health virtues of argan [4]. Exact figures on argan oil exports have been difficult to obtain and in part due to the lack of a specific tariff code, despite indication of growing exports and demand internationally.

Media and marketing campaigns continue to profess the benefits of high-valued export markets for argan, and largely in terms of enhanced livelihoods for rural Amazigh women. A growing tourist economy in the region has also sprung up, and daily tours to women cooperatives producing argan oil are a favorite outing for visitors interested in viewing women huddled together in a room cracking argan nuts and producing argan oil. Professing health and cosmetic virtues, genuine cooperatives, as well as those masquerading to be a cooperative, vie for the business of tourists, whereas marketing agents from overseas engage with federated cooperatives and unions of cooperatives for a stable supply of argan oil for export. More inclined tourists part with their money at a multitude of spas offering relaxing massage treatments with argan oil, some with the conviction that they are improving livelihoods for poor Berber women and the natural argan forest.

Not everyone is, however, convinced of these win-win claims, both domestically and internationally. Lybbert [19], for instance, argues that the argan 'boom' has led to disproportionate benefits for rural households and no appreciable impact on forest health. Well-off households invest in increased goat herds, as well as more aggressive harvesting techniques (harvesting with sticks to knock down fruit). It is argued that these have had significant impacts on degradation, productivity of argan trees, and therefore on incomes from argan oil production [19].

For most rural households in areas adjacent to the forest, key sources of household income are remittances, as well as male-earned income through daily labor employment and sale of argan oil. With rising prices for argan oil, the contribution of argan to household income continues to rise and, at the time of study, was estimated to be on the order of 50%. Field visits and discussions with farmers indicate that prices for argan oil in local markets have risen from approximately 30 dirhams per liter in 2000 to over 150 dirhams in 2014 and as high as 180 dirhams in 2013. Discussions with several cooperative managers indicate that bulk exports of argan oil were within the range of 250 dirhams per liter. Yet, individual (and cooperative) women producers claim that they receive little of this significant margin between the local market price and the export price. One question, therefore, is whether there is any benefit to cooperative production and marketing of argan oil, in terms of exerting greater negotiation power within the value chain.

Figure 1 depicts the argan value chain for private and cooperative producers. One key distinguishing feature of the argan production system is that it has historically been within the domain of women and continues to remain so today. Except for helping to transport the harvest from forest to home, and in marketing argan oil in local markets, men have not been involved in the processing of argan for oil and its joint products. From harvesting semi-dried fruit to peeling, cracking the inner nut to obtain the kernel and hand grinding, one woman will expend close to 3 days of labor in order to obtain 1 L of oil and associated joint products (flesh and paste for animal fodder, shells for sale as heating fuel to bakeries and communal baths). On average, and valued at local market prices, this was equivalent to 80 dirhams (US$10) per day, which was on par with the official minimum agricultural wage rate in rural areas of 75 dirhams per day in 2014.

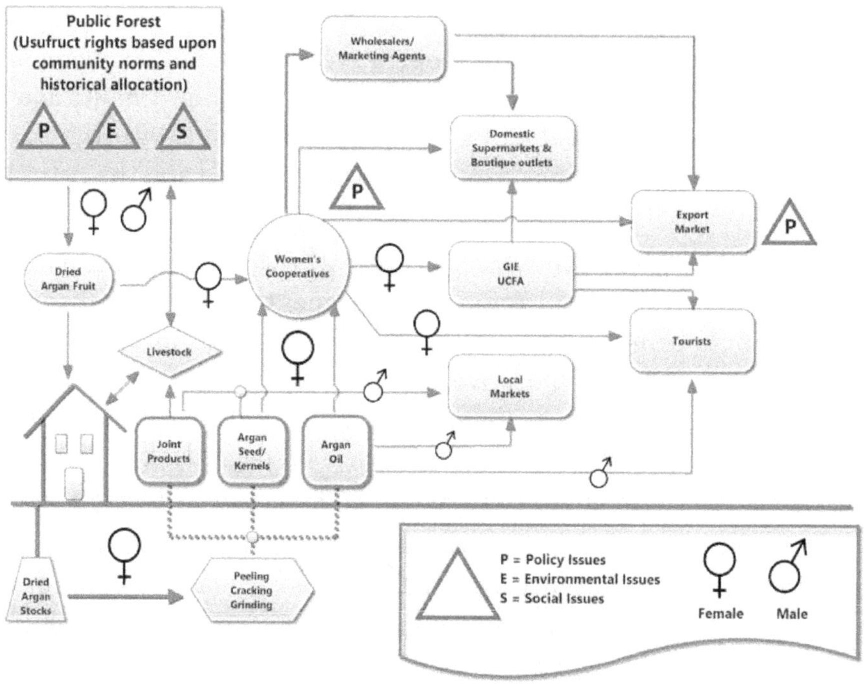

Figure 1.Argan value chain.

While remuneration for labor was at the rate of the official minimum wage, stability in total income generation from argan oil production is a more important issue for the household, not women alone, and this depends on the number of argan trees for which the household possesses rights to harvest. Inherited over generations by male members of the household, usage rights are only acquired by women in the case of no male siblings or on the demise of a spouse. Within the household, therefore, historical norms exist over ownership of income from argan oil sales in local markets, with men taking charge of selling in local markets, retaining the income from sale, and disbursing to household members based on mutual understanding of roles and responsibilities. How many argan trees the household has rights over, and the productivity of those trees in the public forest will, therefore, determine total household income from argan. The lower the number of these rights, the greater the burden is on male members of the household to generate income for household maintenance. A relevant line of enquiry therefore is whether cooperative production and marketing of argan can provide higher incomes, relative to home-based production, and whether this income stream is of a duration which is longer than seasonal home-based production.

At the Taroudent woman's cooperative in Essaouria province, members were paid 45 dirhams for 16 kg of argan fruit brought to the cooperative and placed into the collective pool for peeling, cleaning, and cracking of the outer nut. Piece rate wages were set at 40 dirhams per kg of kernel extracted from the nuts. Based on a norm that each 16 kg of fruit yields 1 kg of kernel, and that one woman in 1 day is able to extract 1 kg of kernels, the average woman member earns 85 dirhams per day of work within the cooperative. This was not significantly different from home-based production when considering transportation costs.

Home-based production is constrained by the number of trees with harvesting rights and by available female labor. For those households with excess female labor in their household, purchasing argan fruit in the local market, over and above the endowment of fruit harvested at the household level, requires a source of cash. Membership in a cooperative does not necessarily help to relieve the constraint of working capital but does help in reducing transaction costs in securing available supply and through pooling of argan fruit between members. Access to a larger pool of argan fruit can come about through the following:

- The sale of argan fruit to the cooperative (either fully or more than what is processed at home) by women cooperative members.
- Purchase of argan fruit from non-members, when hours of manual (collective) labor devoted by women members to the cooperative is more than collective argan fruit brought in by individual members.

Given the nature of poverty and cash constraints, women insist on payment for fruit supplied (at time of delivery) and wages for manual labor after a reasonable period. Both require working capital for the cooperative, and given the lack of access to institutionalized credit within rural communities, it is a necessity for the cooperative to either generate profits or to charge initial membership fees in order to acquire this capital. With relatively similar returns between household production of argan oil and returns to labor at the cooperative, why would women choose to pay fees to join (or maintain) the cooperative? One answer is the benefit of access to a variety of social services provided by the cooperative (child minding, adult literacy lessons, small loans, and ability to socialize). Yet, not all cooperatives provide this service. A more plausible explanation lies in the observation that the production cooperative provides an ability to secure steady wage income over an extended period, with generation of total wage income which far exceeds that from the production of argan oil at home. An added incentive is the ability to retain the value of their labor, away from discretionary use by their husbands or male elders, and for utilization within the household on matters of priority to their needs and the needs of their family. Whether this is a form of empowerment remains moot in so far as it involves trade-offs within the household and between household members.

In taking a more conservative view, mechanical technology for cracking the argan nut has not been developed as of yet. The only available method for obtaining the inner kernel, used for extracting argan oil, is still based on a traditional method of hand and sharpened stone. Women have proprietary rights to this critical production function, based on historical cultural norms related to division of labor, and more specifically to an activity without which the entire value chain for argan oil breaks down. Yet, this advantage is also a potential source of weakness. Cash strapped and resource poor, households relying on argan as a significant source of livelihood need to provide services in order to receive cash, and require immediate cash for services provided. Combined, this results in a situation where women are compelled to work long hours at cooperatives, and at institutions masquerading as cooperatives, for piece rate wages that are on par with the mandated minimum wage. Development of a mechanical cracker will surely put an end to stability in wages from labor currently expended by women in the cracking of nuts, and one of the underlying reasons and benefits for why women currently choose to join a cooperative structure.

With this knowledge, one needs to question a state-sponsored drive for facilitating greater numbers of women's cooperatives under the Morocco Green Plan. It would seem timely for efforts to be concentrated on finding avenues for how rural households can retain control over the argan nut, and to negotiate fair value for either the raw kernel or the oil that accrues from the kernel. Given a need for immediate cash, collective storage units and single selling desks at the community, village, or district level are unlikely to meet with much success unless (i) farmers are paid on delivery and (ii) there is an underlying system for ensuring no leakage, in order to maintain bargaining power with processors and marketing intermediaries. Given the nature of the product, limited options for alternative income sources at the household level, land tenure and property rights issues, the government may need to consider fixing a reasonable minimum price for argan kernel at a rate that (i) fairly remunerates rural households but (ii) still provides an incentive for processors (cooperative or private) to earn a margin on processing and marketing of argan oil.

This potential solution comes at a price. Fixing a minimum price on argan kernel at rates that provide fair remuneration may lead to more aggressive harvesting techniques and potentially negative consequences for forest conservation [19]. Concomitant with any potential consideration of fixing a minimum price on argan nuts, therefore, is the need to consider community-based management of the argan forest. Minimum pricing and forest management must go hand in hand, and without this combined set, options for households to retain income from argan, and for women to continue a cultural tradition of processing argan, are limited in the face of machinery which will ultimately replace hand and stone. The danger here is that without some form of security on the proprietary nature of processing, whether manual or mechanized, the potential for shipping nuts for processing outside of Morocco becomes very real in the face of future innovation.

The story of argan would seem to be one of a race against time and between woman and the machine.

3.2. Rose: Ascent of Hope

Rose production in Morocco has historical linkages to French occupation and largely confined to the village of KalaaM'Gouna in Ouarzazate province. Areas under production and quantities produced have been difficult to assess, but estimates are on the order of 3200 linear km (2 m wide) and production between 2500 and 4000 tons annually [12]. Variation in production is largely due to risks of both frost and sustained drought, but also the nature of the production system itself. Grown in hedge rows and as land boundaries, roses are complementary to principle agricultural production produced on the farm, primarily cereals and summer vegetables, and from which they receive cross fertilization and much of their water requirements.

Pruning, maintenance, and picking have traditionally been undertaken by females in the household, but within the ambit of their main duties in supporting agricultural production of other crops, together with male household members. Women invited to a workshop in the village to discuss challenges and constraints in the production of roses indicated that approximately 10% of their time is spent on roses, mostly during the harvest months of April and May, and with a contribution to household income not exceeding 20%.

Farm households producing roses have several options for marketing (see Figure 2): (i) sale of fresh petals to intermediaries or directly to industrial units, (ii) sale of dried petals to intermediaries for onward sale into national wholesale markets, or (iii) to the cooperative within which they retain membership. In the case of the latter, 10 producer cooperatives now exist, of which 7 are male only and 3 women only. All cooperatives produce the same products—rose water and dried flower petals—with gendered division in roles conditioned by history, culture, and a reaction to current policy.

Dried rose petals and rose water have long been associated with this village, but commercialization of rose petals was only realized in the 1940s when, under French occupation, perfumeries from France encouraged greater cultivation and technical assistance in the cultivation of roses for utilization in the production of oil. Of the two large private industrial units within proximity to the village center, one remains from the French colonial era whereas the other to an investment arm of the ruling Monarch. As detailed by several farmers, and privately by one public extension officer, collusion between the two industrial units is argued to be one of the reasons for low prices paid to growers. Given technical parameters for optimal timing between harvesting and processing of oil (approximately 4 h), farmers are limited in their ability to ship fresh petals to other factories outside of their village or internationally. Perishability, therefore, plays a key role in the ability to collude on prices, as picking occurs at 7 a.m. and delivery to the factory by 10 a.m. in order to be accepted for purchase. Advances in laboratory testing are equally important, therefore, in the ability for firms to collude as approximate time of picking can be verified on delivery.

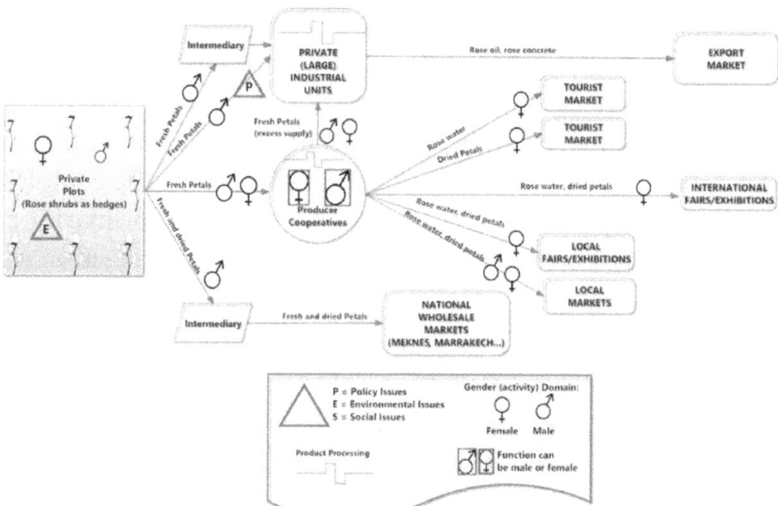

Figure 2. Rose value chain.

With collusion claimed to exist since the 1970s, farmers interviewed detailed how an insurrection in 1990 led to the uprooting of a significant number of rose bushes as a symbol defiance against state control. Sustained bouts of drought and reduction in supply forced firms to raise prices for fresh petals over subsequent years, but the nature of political interference in the setting of prices and control over the industry has not been minimized. Unlike workshops held in other villages for saffron, argan, and cactus, the Ministry of Interior appointed a senior security officer to attend our discussions with farmers, most likely to observe the nature of discussions. Convinced that the workshop was not a threat to the state, or a cause for future civil disobedience, the officer left on the basis of a request and receipt of a letter from the national research institute (INRA), co-hosts of the workshop, and which stated objectives of the workshop. Indeed, many farmers invited to the workshop did not attend, and on follow up, indicated that they were fearful of repercussions from discussions related to pricing and setting of prices for fresh rose petals.

The political nature of rose production systems, and currently limited contribution to household income, raises interesting questions of why rose has been included within the Morocco Green Plan. With the highest estimated production of fresh roses at 4000 tons annually, and a price of 13 dirhams per kg for fresh rose petals, small farm households within the village stand on a potential income gain of 52 million dirhams (approximately US$6.5 million) annually. This estimated income gain is based on potential market supply, given that only 30% of production is currently sold in fresh form to industrial plants engaged in the production of oil, 3% to cooperatives engaged in the processing of rose water, and the remainder (67%) sold in dry form. With a conversion ratio of 4:1 of fresh to dried, and a (high) price of 100 dirhams per kg for dried roses, estimated potential income for farm households from dried rose marketing was 6.7 million dirhams (approximately US$850,000) and 17.2 million dirhams (approximately US$2.1 million) from fresh rose sales to industrial units and cooperatives.

In light of these figures, and given Morocco's ranking as third largest producer of rose oil [6], a concentration on roses under the Morocco Green Plan is valid. Yet, the formation of cooperatives under Pillar II of the plan and poverty alleviation through production cooperatives is somewhat difficult to understand for the rose sector. On the basis of focus group discussions, there is a generally held notion that (i) it is easier to obtain grants and subsidies for women cooperatives, (ii) for the purpose of marketing dried roses and rose water, the characteristic of the product, and the target consumer, lends itself to better sales if marketed by women, and (iii) applications for the start-up of cooperatives by the youth (males in particular) are sometimes viewed as threatening to local security services and face long delays. Women appear to be less threatening as an organized group of producers relative to males and specifically young males.

Returning to the calculation on potential income from the sale of fresh and dried roses, a focus on expanding marketing and sale of dried roses is clearly in the interest of farmers given (i) greater choice in markets, (ii) an ability to diminish the constraint of perishability and more importantly, (iii) removing surplus of fresh petals from the local market in order to influence a higher price paid by industrial processors. While rose water production from fresh petals, at the cooperative level, offers an interesting option for diversification, relatively small scales of production, high equipment costs, and competition from synthetic imports (rose essence) limits the ability for take-off and impact for rural livelihoods. Equally important is a move toward organic certification, particularly for products that are within the wider rubric of 'cosmetic' products. But are farmers within KalaaM'Gouna able to respond to this niche? Planted as hedges around small-scale production areas of cereals and vegetables, cross contamination of fertilizers and pesticides is impossible to avoid within the current production paradigm practiced, and therefore, the ability to fill this niche is limited.

Is there a possibility to expand areas of production under rose, and with new paradigms of production that have sound environmental underpinnings, but still commercial in nature? Farmers responded positively to this possibility, but noted a fundamental constraint in access to productive land that has been in reserve since French colonization and allocated to senior military officials under favorable lease. Inability to access previous tribal lands that were nationalized at independence is also a source of concern and limits potential for rural households to engage in more commercialized rose farming systems. Farmers also mentioned that in an environment where there is excess supply of fresh rose petals in the local market, with no options for exporting fresh petals outside of the village, commercialized production by farmers will only lead to further downward pressure on fresh petal prices. While relevant and correct, the fact that a recent private entrant into the industrialized sector has initiated production of rose oil, with purchase of fresh rose petals from the village, suggests that there is scope for industrial expansion of rose oil, and therefore a larger market for the sale of fresh rose petals.

That the new entrant has not engaged in collusion on prices with existing industrial units is a positive development, but there is one looming risk. Farmers and extension agents state that the new industrial unit has also initiated production of roses on a commercial scale within the village and to supply its own plant. Scale of operation will determine whether this commercial production of roses displaces any current supply from small households to the other two industrial units, and what impact this will have upon producer prices for fresh rose petals. On a positive note, larger scale of production may result in wage opportunities for picking, rates for which will depend upon availability of specialized female labor in season and at critical periods.

So, we return to the question of whether small-scale farmers should be concentrating on expanding cooperative production of rose water or concentrating on dried roses. Market studies are clearly needed to answer this question, but cursory analysis presented herein suggests that the government should revisit the premise for promoting producer cooperatives, particularly for women, without a clear understanding of available markets for the product under transformation, the specific nature of the product itself, and constraints of expanding production of primary inputs for supply to the cooperative. Moreover, if the government is indeed interfering in the setting of prices, through collusion between Monarch investment units and private industrial units, the premise of Pillar II of the Morocco Green Plan is seriously undermined for the rose sector.

Small rose farmers in the rural village of KalaaM'Gouna continue to rely on remittance incomes, limited production of cereals, and on a hope that roses will one day pave the path out of poverty.

3.3. Cactus: Waiting for a Ride

Cactus production in southwestern Morocco is predominantly undertaken within the province of Sidi Ifni on: (i) small-scale private land, (ii) marginal public land, and (iii) on hills of marginal quality land. With an average of 126 mm of precipitation, cropping options are limited, and proximity to the ocean provides an ambient environment for cactus to prosper. Yet, untimely precipitation or poor distribution of precipitation is a significant risk to producers. In 2013, farmers and industry representative interviews suggest that poor and uneven rainfall was the cause of loss for 80% of harvestable yields, well above the typically high loss of 40–50% in an average year. Discussions with farmers and public extension agents indicate that the severity of loss is related to an inability for producers to ship to market in a timely manner, and largely due to a lack of transportation and marketing intermediaries at critical points of the season. While markets in Agadir-Inezgane and farther north place a premium on cactus from Sidi Ifni, market intermediaries choose their routes based upon profitability and time of transport. Tomatoes, citrus, and berries produced in areas of proximity to Agadir are naturally more lucrative and, therefore, number of vehicles and middlemen travelling to the southern region of Sidi Ifni during vegetable and citrus harvesting seasons are argued to be in limited supply.

With claims of limited transportation options and cactus rotting in the field at such high rates, one is perplexed by recent initiatives under the Morocco Green Plan to subsidize the planting of cactus shrubs at a cost of 4000 (approximately US$500) dirhams per hectare. One explanation offered by officers at the Ministry of Agriculture is that the Morocco Green Plan has set a target of reducing the area under rainfed cereal plantation, and given limited options for replacement in an area with limited precipitation, cactus is a natural choice. In addition, the plan also sets targets for limiting desertification and replanting of public forest lands, an equally compelling reason for pushing cactus plantation into areas of marginal land productivity.

Discussions with farmers indicate that an average household is likely to earn 25,000 dirhams per year from the sale of cactus, which is lower than the minimum agricultural wage rate. Given that there are limited alternative cropping options for most small rural households in the area, men have traditionally engaged in apiculture, whereas women devote their labor to harvesting and basic maintenance of cactus plantations. Considered a lazy man's' crop, rain-fed production requires little maintenance and no material inputs in the production of cactus fruit. All sales of fresh fruit from the farm gate are therefore combined returns to female labor (including management) and returns to male contribution in their traditional role of marketing.

Given issues of perishability and need to immediately ship to market at harvest, fresh fruit cooperatives, outside of larger urban markets, are unlikely to succeed, unless they add further value to the product. In attempting to mitigate this constraint, the current cooperative structures in cactus are comprised of (i) male cooperatives, which are merely organizational structures for maintaining collection centers/staging areas which organize collective member fruit for shipment to federated cooperative structures (GIE) and onwards for export, (ii) women cooperatives that extract cactus oil and prepare a variety of dried and processed cactus products.

Figure 3 illustrates the value chain for cactus fruit in the Sidi Ifni province. Exports of fresh fruit through the GIE have been limited and undertaken from a recently built factory provided under the Green Plan. At a cost of 12 million dirhams (approximately US$1.5 million) a visit to the factory reveals lavish office spaces, packaging equipment, cooling facilities and all necessary warehouse equipment, but no revolving working capital for salaries, purchase of packaging material, and most importantly payment for fresh fruit supplied by farmers within the region. A one-time grant for testing the operation, with a shipment to Eastern Canada earlier this year resulted in complete loss on arrival and due to a lack of appropriate cooling and packing measures on departure from Morocco.

This initial export experience has highlighted fundamental issues of limited coordination between various public and private actors, and a potential weakness of the Green Plan. A lot of effort has seemingly been placed on infrastructure development and outputs rather than outcomes and investments in approaches that are sustainable, replicable and with measurable impact on small farming households—the intended beneficiaries of pillar II of the Green Plan. In the case of limited numbers of women cactus cooperatives, subsidized equipment for extracting cactus oil, renovation of buildings and small equipment for manual processing of cactus (pickled cactus ears, dried fruit, etc.) are being undertaken by the Morocco Green Plan as well as other international partners such as the United Nations Development Programme (UNDP).

On the positive side, a stable market for cactus oil has the potential to mop up significant amounts of cactus fruit that lays rotting in the fields, only a portion of which is fed to the limited number of livestock within the immediate area. With 30 kg of fresh fruit required to obtain 1 kg of seed, and between 20 and 60 kg of seed to produce 1 L of oil, each liter of cactus oil has the potential to remove between 600 and 1800 kg of fresh fruit from the field. At current retail prices of close to US$500 per liter, a stable market for cactus oil has the potential for a win-win solution if there is limited variation in the range of seeds needed to produce oil. At the high end of 1800 kg, the product is unprofitable.

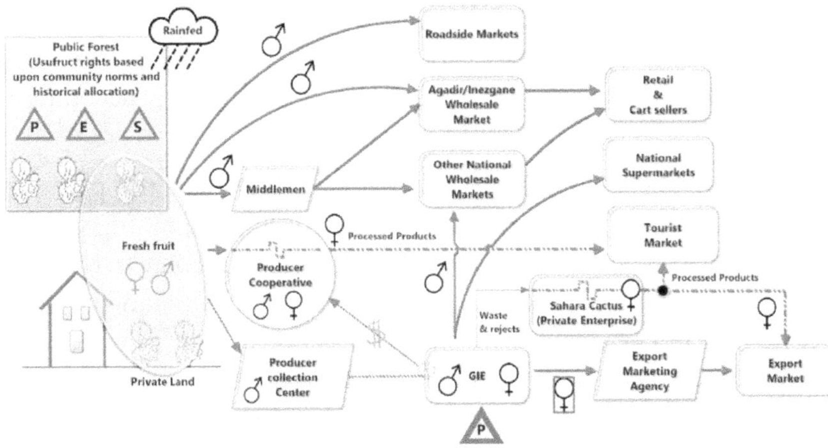

Figure 3. Cactus value chain.

The fundamental problem with variation in seed quality for oil production lies in rainfed production of cactus. Effective productivity and product for processing requires commercial production of cactus, with provision of drip irrigation and application of judicious amounts of management and technical inputs. Commercial trials in this regard have been initiated by private entrepreneurs in Sidi Ifni and with plantings to bear fruit this year. Whether or not this production is tied to a specific industrial unit for production of processed cactus products is unclear. What the investment does suggest is that there are perceptions of profit to be made in cactus, but whether or not this results in trickle down benefits to small household producers is less clear. While market development for processed cactus products, particularly oil, will have spill over benefits to the local community in terms of wage labor, it is unlikely to have an impact upon the volume of sales for fresh and ripened cactus at the farm gate for small producers. This is largely due to variation in quality of seed for processing and due to climatic variation and the need for standardized commercial production, which has now been undertaken.

For small producers of fresh cactus fruit, the only significant opportunity lies in sales of fresh fruit to national and international markets. The argument that middlemen do not purchase cactus in high volumes during the tomato and citrus seasons is somewhat weak. Based on discussions with wholesalers and street cart vendors in the urban center of Agadir, a more likely explanation is an economic one, and that related to the protection of marketing margins through not flooding local wholesale markets. A dedicated study is required to study this conjecture, but one underlying question is why cactus farmers do not collectively hire trucks and ship their cactus to market, in the absence of the middle-men. Access to credit is brought up frequently in this regard, but is also a weak argument given that sales of cactus in wholesale markets are paid in cash on delivery. With perseverance, a more plausible explanation of cactus under the domain of women in the household, and men tending to their bees in the hills and valleys was uncovered. Without significant male support, women are reluctant to engage in the marketing of cactus given cultural norms and particularly so if this requires their accompanying the shipment to Agadir in order to secure the sale.

This social (and perhaps) cultural constraint brings back into play the export-oriented factory built under the aegis of the Green Plan, and an opportunity for this facility to play a critical role in providing significant income to the region. Why there has been limited movement in supplying necessary skill and working capital to leverage investments already undertaken is a source of confusion to many within the cactus sector. What is clear is that there is a lack of coordination within the implementation of the Morocco Green Plan, and specifically between a number of governmental agencies with mandates to provide (sometimes competing) support services.

Cactus production in Sidi Ifni is predominantly under the domain of women who harvest, sort and prepare cactus for sale at the farm gate. Unable to take the ride to market, in order to sell their fruit during periods of excess supply, women have been relegated into the passenger seat and are in transit awaiting a ride to ship their cactus products to market.

3.4. Saffron: All We Need is Cash

With a relatively small population of 12,000, the village of Taliouine sits on the road between the commercial center of Agadir and the rose-producing village of KalaaM'Gouna. With over 2000 families involved in the production of this crop, the importance of saffron to stability in livelihoods, and to the economy of the village is significant. Male and female roles in the production of this highly prized commodity internationally are delineated along lines of comparative advantage. The application of brute strength is provided by men in the preparation of fields for planting and water, while women engage in early morning picking of flowers, and a delicate procedure of plucking stamens from the flower.

Labor shortages, particularly for skilled female labor, are implicit in norms related to payment for picking, with one-fifth of the output payable in a year when labor is abundant, and up to one-half when skilled female labor is in short supply. Focus group discussions with women suggest that a historical culture of self-help (Tiwizi), where women collectively harvested and assisted each other in the plucking of stamens has eroded over time. This is largely due to emigration of male household members, leaving behind a greater burden of household duties for women; as well as general societal shift toward individuality. Interestingly, male cooperative members suggested that women cooperatives for saffron were an avenue through which to revive a historical custom of Tiwizi, a claim dismissed by female cooperative members in a separate discussion. Women suggested that a primary reason for women cooperatives to form was to obtain free saffron bulbs, provided by the public extension services, and as a programme aimed at fostering expansion of saffron production under the Morocco Green Plan. While interesting, only 2 of the 36 existing cooperatives in the village are women only, and of the remaining 34, only 2 are mixed cooperatives. In the case of the latter, women belonging to this cooperative were widows or women whose husbands were away from the village for income-generating opportunities. A heavy concentration of male only cooperatives is clear indication of the strong role that males play in the marketing of saffron and in decisions related to the planting of saffron.

With up to 50% of land planted under saffron for an average household (ranging from ½ ha to 1 ha), the decision of land allocation is based on labor and water constraints. Given small parcels of land and limited family labor, households interviewed indicated that saffron will always remain a family-based activity for the household and not a commercial activity. In large part, this also reflects the role of saffron in smoothing out income during the months of November to January, when there is limited income from agricultural activities. A claim that saffron has never been immensely profitable to the farm household is plausible, given an understanding of income streams over the year and current challenges of credit within the sector. Traditionally, and even contemporarily, saffron continues to be used as a form of currency within the village, in barter trade, as well as in setting up accounts with shopkeepers for payment by saffron at the time of harvest. This continued reliance on saffron as currency in hard times raises a larger issue of market power and lack of ability for farmers to negotiate better prices.

The role of cooperative formation and social organization under the Morocco Green Plan was aimed at improving farm margins and access to markets for local products, with saffron identified as one key product. Yet, exports through cooperatives remain limited, with most sales of saffron still destined for local and national markets. In limited cases, and with the support of internationally sponsored NGOs, one male cooperative has been successful in the supply of saffron to Italy on renewable yearly contracts. Others have informal contacts with overseas buyers who place orders sporadically, with prepayment through the post office and shipment through courier. In general, cooperatives rely on tourist flows, sales at exhibitions and fairs, and to middlemen who onward sell into national markets.

Figure 4 illustrates the value chain for saffron in Taliouine village. One interesting point to note is the sale of saffron to the GIE (groupementd'intérêtéconomique). As a second level cooperative promoted by the Morocco Green Plan, its objective is to act as a single selling desk for cooperatives, in order to negotiate better prices and to promote exports. While interesting in theory, a significant challenge in access to working capital renders the GIE as an organization of little value in practice.

Due to cash constraints, the GIE is unable to pay saffron farmers on delivery. As in the case of argan, poverty and a need for cash mitigates the ability for poor households to sell saffron to the GIE on consignment. With 300 kg of storage capacity and an average market price of 13 dirhams per gram, the GIE would require 4 million dirhams of revolving credit (approximately $US500, 000). Management of the GIE suggest that while high, it is possible to obtain credit from formal banking institutions, but a requirement for personal guarantees from the directors of the GIE, as well as collateral for securing the loan mitigate this potential. A lack of coordination and missing links within the implementation of the Morocco Green Plan comes up again very strongly. One (male) farmer within a focus group discussion asked the question of why the government engages in the supply of free saffron bulbs to farmers, but does not buy the saffron from farmers at reasonable prices. The GIE is asking a question of why the Morocco Green Plan has implemented a model that it does not want to fund in order to get it off the ground.

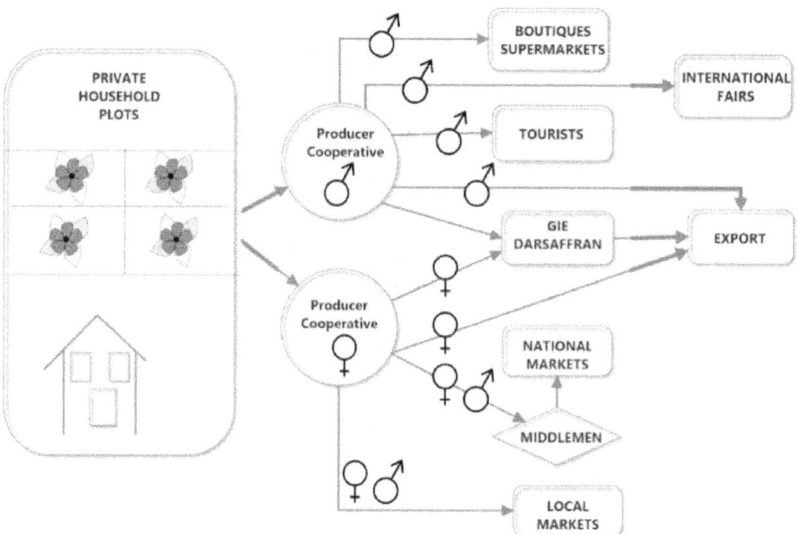

Figure 4. Saffron value chain.

Amid all of this confusion, traders and market intermediaries buy saffron from the local market at prices close to the cost of production and sell onwards to national markets or international buyers at a margin of more than 10 dirhams per gram. With estimated production of over 3000 kg per year in Taliouine, possession of cash translates into market power, and at present, power does not lie in the hands of the cooperative or the farm household due to a lack of cash.

4. REFLECTIONS

Given our focus on the gendered implications of pillar II of the Morocco Green Plan, our unit of analysis has been the agricultural cooperative. Cooperative production, processing, and marketing are being promoted, given conventional wisdom that small farmers are disorganized in production and marketing and, therefore, subject to opportunistic behavior of marketing agents—both Moroccan and international. Focus group discussions and key informant interviews indicate a farmer-centric view that disorganization at the landscape level reflects a lack of coordination between government agencies, where roles and responsibilities of public bodies are either not clearly defined or overlapping. Traditional transfer of technology models within the system of public extension continue to persist, with little (ostensible) movement toward more participatory approaches for innovation that embody joint learning and action-oriented research. Subsidies on primary inputs (seed, fertilizer, fuel) and public investment in infrastructure (equipment and buildings) for agricultural cooperatives are provided without matching investments in the provision of technical training, for knowledge on effective practices related to storage and transportation of perishable commodities to national and international markets, as well as in the provision of knowledge on effective management practices. In those limited instances where technical assistance is provided and gratefully acknowledged by producers, there are claims of interference by the Ministry of Interior in price fixing (such as for rose petals). This, in collaboration with parastatal processing plants, such that the focus for many small farmers is on quantity and not necessarily quality. We are unable to validate the claims of price fixing, though anecdotal evidence does point to the same. If found to be true, the Green Plan objective of enhancing access to national and international markets is severely undermined.

Our findings also reveal that constraints on land, labor, and equitable access to working capital, exacerbated by the impact of sustained drought on productivity potential, are crosscutting for all four value chains studied. While these issues can be addressed over-time through appropriate sectoral and cross-sectoral interventions, it is the risk of elite takeover within cooperative structures that presents a danger for further entrenchment of rural poverty. More effective organization of production for the commodities studied, each with its own unique specificity of production and demand, may ironically lend itself to invited commercialization and thereby capture of cooperative production units—away from cooperative decision making by its members and toward private vested interests. How commercialization can be managed, without exclusion of the marginalized (particularly women) will depend on how the government eventually comes to terms with many difficult issues:

- One premise of the Green Plan is a desire to move away from a history of paternalism through the subsidization of agricultural production and state-led directives for cooperative production. Relaxing the boundaries around which civil society organizations are permitted to operate would be in keeping with the tenets of the plan; yet recent geopolitical events within the region ("Arab Spring") may be of significant worry for national security systems. How civil society organizations in rural areas (cooperatives, community-based organizations, non-governmental organizations, self-help groups, etc.) achieve greater independence in terms of organizational structures and economic independence over time remains an open question. Within the current environment of regional insurrection and instability, it is quite possible for paternalism to be strengthened as a counter measure to potential insurrection.
- In arguing for greater independence of civil society organizations to operate, for the enhancement of well-being within rural communities, there is a clear need for the state to tie the Green Plan with a wider process for area development—not simply agricultural development. While value chain initiatives, undertaken on a commodity basis, are necessary and important in terms of improving livelihoods, trade-offs between economic empowerment and well-being are likely to lead to potential disempowerment of marginalized members of society. Our analysis reveals that many of the women within the cooperatives visited work significantly long hours (often 12 or more hours in a day), with limited opportunities for institutionalized child care. While reports exist of increasing enrolment of girls into high school [19], given improved economic circumstances, our analysis indicates that rural families may have no choice but to keep their daughters at home. Despite subsidies on school fees and options for boarding, our discussions revealed that in many cases, school enrolment in peri-urban areas is not sustained given high transportation costs for children to return home for the weekend or for national holidays. Family dynamics and livelihood structures have ostensibly changed with the proliferation of production cooperatives. Young mothers are more than likely to have their adolescent children accompany them to the cooperative. Some cooperatives provide child minding and educational services, but most are basic in terms of infrastructure and their ability to provide social services. With larger families, the burden of child care often, therefore, befalls on older siblings—sisters. One trade-off from an economically empowered mother, through more profitable and equitable interventions within value chains, is a potentially disempowered daughter, who may be deprived from sustained educational activity, and equally important, the ability to learn through play.

- Within our focus group discussions (male, female, mixed groups), there was consensus that the household, as a unit, was much better off economically since the proliferation of production cooperatives. Males clearly acknowledged changes to social and cultural norms related to the participation of women in nontraditional economic activities. Females felt more independence and a sense of empowerment given greater respected accorded by males within their household, and within their communities more generally. When asked if this is a fundamental change in societal norms, and based on self-realization, there was much concern related to long-term embedded change. Males appeared to be content with the realization that increased earnings through economic engagement of female members within the household is an increase in their own (gendered) disposable income. Increased female earnings translate into reduced burden for males to cover household expenses, inclusive of child rearing, educational and medical expenses. With the risk of potential unraveling of economic opportunities, for example within the argan sector, will restructured social and cultural norms for female economic engagement persist, or will they revert to historical constraints?
- The existing Green Plan implementation structure, which relies on paternalistic and subsidized programs, allows market intermediaries to exploit disparities and marginalization to gain market power. The existing literature on agricultural innovation predominantly emphasizes value chain development initiatives. Although enhancing marketing margins by improving connections between producers and other stakeholders in the value chain is important, there is a lack of focus on the compromises that arise at the home and community level due to innovation. Trade-offs occur in various areas, including health, nutrition, education, farm and off-farm income opportunities, and notably in terms of unequal gender-based access to resources and economic opportunities for employment. Emphasizing innovation systems enhances efforts to enhance efficiencies in value chains by incorporating insights into the process of change and promoting more mutually beneficial and sustainable social and economic relationships among stakeholders. Innovation systems are social systems consisting of actors united by a shared objective of addressing common challenges and constraints of mutual concern. These systems exist naturally but need facilitation and coordination to enhance efficiency, equity, and equal opportunity in accessing resources, including knowledge resources. Donor-funded project-based efforts are unlikely to successfully establish effective processes for innovation within and between value chains, as well as through connections with broader area development strategies. This is

mainly because of present geopolitical issues that are expected to restrict the number and scope of firms authorized to carry out such activities, as well as project schedules that are short in duration. International and national research agencies are well-positioned to promote innovation systems that prioritize inclusivity and equity. It is uncertain if they are prepared to handle this duty, especially when it comes to addressing potentially delicate topics of social policy change and empowering rural communities.

5. AT A GLANCE

The Morocco Green Plan (MGP) has provided substantial economic advantages to small agricultural households. By focusing on enhancing efficiency and profitability in value chains for important local commodities, the establishment of women's cooperatives has resulted in beneficial advancements in female empowerment. Our study used qualitative and participatory research methodologies to examine how gender influences value chains for argan, rose, cactus, and saffron in southern Morocco. We found that economic empowerment, facilitated by women's cooperatives, is precarious and vulnerable to many risks. This is mainly due to a state-driven strategy that has not adequately addressed the inequalities present in value chains for important local products, and the integration of current social and cultural norms with the principles of a national effort to 'modernize' the agricultural industry. We recommend that the MGP is impartial regarding gender in this regard. Embedding value chain improvement efforts in an innovation systems framework rather than a state-centric approach is more likely to promote prosperity in rural communities, along with sustainable social and economic benefits in value chains that serve the disadvantaged.

REFERENCES

1. Badraoui M, Dahan R. The Green Morocco Plan in relation to food security and climate change in dry areas. In: Proceedings of the International Conference on Food Security and Climate Change in Dry Areas; 1-4 February; Amman, Jordan. ICARDA; 2010
2. Kassam S. Reaching out to poor and marginalized communities, learning alliances for out-scaling value chain innovations in the MENA region. Workshop Report. Part 1, June 3-4. Agadir, Morocco: ICARDA; 2013
3. Kassam S. Reaching out to poor and marginalized communities, learning alliances for out-scaling value chain innovations in the MENA region. Workshop Report. Part 2, September 16-17. Agadir, Morocco. ICARDA; 2013
4. Farines M, Charrouf Z, Soulier J. The sterols of Argania spinosa seed oil. Phytochemistry. 1981;20:2038-2039

5. Crocusbank Project [Internet]. 2013. Available from: http://www.crocusbank.org/Index18En.htm [Accessed: October 25, 2013]
6. Kovacheva N, Rusanov K, Atanassov I. Industrial cultivation of oil bearing rose and rose oil production in Bulgaria during the 21st century: Directions and challenges. Biotechnology & Biotechnological Equipment. 2010;24(2):1793-1798
7. Prendergast HDV, Walker CC. The argan: Multipurpose tree of Morocco. Curtis's Botanical Magazine. 1992;9(2):75-85
8. El Aich N, El Assouli N, Fathi A, Morand-Fehr P, Bourbouze A. Ingestive behavior of goats grazing in the Southwestern Argan (Argania spinosa) forest of Morocco. Small Ruminant Research. 2007;70(2):248-256
9. Miloudi H, Charrouf Z, Soulhi A, Hachimi L, Guillaume D. Influence of origin and extraction method on argan oil physico-chemical characteristics and composition. Journal of Agricultural and Food Chemistry. 2005;53(6):2081-2087
10. Charrouf Z, Guillaume D. Second metabolites from Argania spinosa (L.) Skeel. Photochemistry Reviews. 2002;1:345-354
11. Ait Oubahou A, El Otmani M. La culture du safran. Fiche technique. Transfert de Technologieen Agriculture. 2002: Bulletin mensuel du PNTTA n°91, 4p
12. Karra Y. Contribution au développementagronomique des produits de terroir de la région de Souss Massa Draa. Cas de la rose à parfum de KelaâtM'gouna et du Safran de Talouine. Rapport de fin de stage en guise de titularisation au poste de Chargé de Recherche Grade « A » enAgronomiegénérale. Agadir: INRA; 2011
13. Lybbert TJ, Barrett C, Narjisse H. Market-based conservation and local benefits: The case of argan oil in Morocco. Ecological Economics. 2002;41(1):125-144
14. Lybbert TJ, Barrett C, Narjisse H. Does resource commercialization induce local conservation? A cautionary tale from southwestern Morocco. Society & Natural Resources. 2004;17(5):413-430
15. Lybbert TJ. Patent disclosure requirements and benefit sharing: A counterfactual case of Morocco's argan oil. Ecological Economics. 2007;64(1):12-18
16. Biermayr-Jenzano P, Kassam SN, Aw-Hassan A. Understanding gender and poverty dimensions of high value agricultural commodity chains in the Souss-Masaa-Draa region of southwestern Morocco. Working paper. ICARDA; 2014
17. M'hirit O, Benzyane M, Benchekroun F, El Yousfi SM, Bendaanoun M. L'Arganier: Une especefruitiere-forestiere a usages multiples. Available from: ftp://ftp.fao.org/docrep/fao/009/ah001f/AH001F01.pdf [Accessed: October 15, 2013]

18. Gonet W. Avancement des travaux de recherche sur l'extraction de l'huiled'argan au niveau du ménage. Journéesd'études sur l'Arganier. 23-24 Juin. Essaouaria, Morocco. 1998

19. Lybbert TJ, Magnan N, Aboudrare A. Household and local forest impacts of Morocco's argan oil bonanza. Environment and Development Economics. 2010;15(4):439-464

FOOTNOTE

Chapter 8

1. We use "world market" and "international market" interchangeably in this article.
2. For the simplicity we keep the term world market price for describing the long-run price equilibrium.

Chapter 10

1. One of the authors of the current paper participated in the mentioned FAO project and the chapter is based on the respective FAO Report titled "Processor-driven integration of small farmers into value chains in Eastern Europe and Central Asia (EECA)" [9].

Chapter 12

1. For example, World Bank (WB), African Development Bank (AfDB), United States Agency for International Development (USAID), Department for International Development (DFID), International Development Research Centre (IDRC), Trust Africa, European Union, and African Union

Chapter 13

1. A recent review on these issues is provided by Santeramo and Di Gioia [9].
2. The adjustment costs are defined as costs associated with changing retail prices and subsequently adapting retail logistics, wholesale costs and sales (e.g., advertisement and relabeling costs, storage and volume discounts).
3. A detailed review of major findings is provided in Table 1.

INDEX

www.ingramcontent.com/pod-product-compliance
Lightning Source LLC
LaVergne TN
LVHW062118200926
848484LV00007B/1705
* 9 7 8 1 7 8 7 1 5 2 7 9 3 *